emerald PUBLISHING

ice Publishing

AF621440

ICE Core Concepts: Low Carbon Cements and Concrete for Construction

Professor Monower Sadique CEng FICE

Published by Emerald Publishing Limited, Floor 5, Northspring, 21-23 Wellington Street, Leeds LS1 4DL.

ICE Publishing is an imprint of Emerald Publishing Limited

Other ICE Publishing titles:
Appraisal and Repair of Existing Concrete Structures
Yue Choong Kog. ISBN 978-0-72776-693-9
Cement Chemistry: Calcium silicates and anhydrous Portland cement
Ian G Richardson and HFW Taylor. ISBN 978-1-83549-777-7
ICE Handbook of Concrete Durability
Edited by Marios Soutsos. ISBN 978-0-72776-375-4

A catalogue record for this book is available from the British Library

ISBN 978-1-83608-477-8

Cover photo: courtesy of National Grid

Commissioning Editor: Michael Fenton
Content Development Editor: Cathy Sellars
Books Production Lead: Benn Linfield

Typeset by: KnowledgeWorks Global Limited
Index created by: Lyn Nesbitt-Smith
Printed and bound by CPI Group (UK) Ltd, Croydon, CR0 4YY

Dedication

To the loving memories of my mother and father.

To Afrina, for her unreserved support.

To Sabique and Nahla, for their encouragement.

Contents

Acknowledgements

I extend my heartfelt gratitude to my beloved wife, Afrina, my wonderful son, Sabique, and my sweet daughter, Nahla, for their unwavering support and encouragement throughout the writing of this book. Their love and understanding made this journey possible. I also wish to express my deep appreciation of my colleagues at Liverpool John Moores University, whose invaluable support and knowledge have greatly enriched my work.

Permission to reproduce extracts from British Standards is granted by BSI Standards Limited (BSI).

Extracts from Global Cement and Concrete Association (GCCA) documents and reports are reproduced with permission of the Global Cement and Concrete Association (GCCA).

Extracts from *Guide to Concrete Construction* are reproduced with permission of Cement Concrete and Aggregates Australia (CCAA).

Extracts from *Concrete Information – Types and Causes of Concrete Deterioration* are reproduced with permission of Portland Cement Association (PCA).

Merseyside Fire and Rescue Service (MFRS) have granted permission to use images.

The British Cementitious Paving Association (Britpave) have granted permission to use an image.

National Grid have granted permission to use an image.

About the author

Professor Monower Sadique CEng FICE is Professor of Construction Materials at the School of Civil Engineering and Built Environment, Liverpool John Moores University, UK. Professor Sadique is a Fellow of Institution of Civil Engineers (ICE). He earned his first degree in Civil Engineering from Chittagong University of Engineering and Technology (CUET), Bangladesh. He began his professional journey in 1999 by joining the Bangladesh Civil Service as an Assistant Engineer in the Roads and Highways Department (RHD) under the Ministry of Communication. During his tenure, Professor Sadique contributed to the construction of various concrete girder bridges at national and regional highways across Bangladesh.

In 2009, he achieved an MSc in Road Management and Engineering with Distinction from the University of Birmingham. He went on to complete his PhD in 2012 at Liverpool John Moores University (LJMU), where his research titled 'Development of Low Carbon Cement from Waste Biomass Fly Ash' laid the foundation for his career in sustainable construction materials. In 2017, Professor Sadique joined LJMU as a Senior Lecturer, progressing to the rank of Professor due to his outstanding contributions to research, teaching, and industry collaboration.

Professor Sadique's commitment to low-carbon construction materials is rooted in his unique career trajectory, which bridges practical industry experience and academic research. His transition from a highway engineer to an academic researcher has equipped him with a deep understanding of industry challenges and the scientific expertise required to develop innovative solutions.

Over the past decade, Professor Sadique has been at the forefront of pioneering research in industrial and academic settings. His work spans the UK highways sector, collaborating with public and private organisations, Small and Medium-sized Enterprises (SME), universities, and manufacturers. Professor Sadique's diverse research portfolio includes cement and concrete innovation, low-carbon construction materials, highway design, and asset management. Within ten years of earning his PhD, he secured four patents, supervised six PhD projects, published over 50 journal papers, and generated external grant funding – all focused on addressing critical challenges in low-carbon construction.

Driven by a passion for transformative knowledge exchange, Professor Sadique is dedicated to creating practical and scalable solutions that address material security challenges posed by climate change. The publication of this book exemplifies his strategic approach to advancing research and knowledge exchange. It aims to equip both aspiring and practicing engineers with essential and up-to-date knowledge in the civil, construction, and materials sectors, fostering sustainable practices for the built environment.

Table 1.1 Mechanical and physical requirements set out in BS EN 197-1 (BSI, 2019a) for conformity of common cements

Strength class	Compressive strength: MPa Based on BS EN 196-1 (BSI, 2020)				Initial setting time: min Based on BS EN 196-3 (BSI, 2016a)	Soundness: mm Based on BS EN 196-3 (BSI, 2016a)
	Early strength		Standard strength			
	2 days	7 days	28 days			
32.5L	–	≥12.0	≥32.5	≤52.5	≥75	≤10
32.5N	–	≥16.0				
32.5R	≥10.0	–				
42.5L	–	≥16.0	≥42.5	≤62.5	≥60	
42.5N	≥10.0	–				
42.5R	≥20.0	–				
52.5L	≥10.0	–	≥52.5	–	≥45	
52.5N	≥20.0	–				
52.5R	≥30.0	–				

Table 1.2 List of permitted second main constituent and their notations specified in BS EN 197-1 (BSI, 2019a)

Permitted additions	Notation
Granulated blast furnace slag	S
Natural pozzolana	P
Natural calcined pozzolana	Q
Siliceous fly ash	V
Calcareous fly ash	W
Burnt shale	T
Limestone	L
Silica fume	D

by a 40 × 40 × 160 mm prism of mortar after 28 days under water curing. There are three strength classes: 32.5, 42.5, and 52.5; cement needs to achieve equal or more than the specified strength after 28 days according to BS EN 196-1 (BSI, 2020). The prism of mortar should be made of a cement to fine aggregate (sand) to water ratio of 1:3:0.5 by weight. This number in the nomenclature is then followed by a letter: either N (ordinary early strength), R (high early strength) or L (low early strength). This nomenclature is summarised in Figure 1.2.

For example, a cement denoted EN 197-1 CEM I 32.5 R is Portland cement conforming to EN 197-1, of strength class 32.5 with high early strength.

Figure 1.2 Cement designations/nomenclature proposed in BS EN 197-1 (BSI, 2019a)

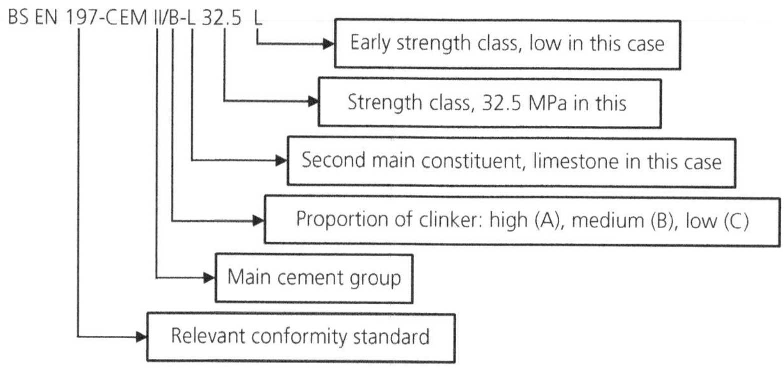

A cement denoted EN 197-1 CEM III/B-S 42.5 N is blast furnace cement conforming to EN 197-1, containing between 66% and 80% by mass of granulated blast furnace slag (S), of strength class 42.5 with ordinary early strength.

According to BS EN 196-1 (BSI, 2020), strength conformity for any cement or cementitious binder needs to be conducted using 40 × 40 × 160 mm prism mortar; details of the test are discussed in Chapter 5.

Emission issues of Portland cement production

It is believed that concrete, which is a composite material, is the world's second most consumed substance after water. Depending on the intended use and mix design, in typical concrete (by mass), 70–85% is taken by aggregates (coarse and fine), 10–20% by cement or cementitious binder, and the rest, around 3–6%, by water. Cement is termed as the glue of progress as it is the fundamental material for building society's infrastructures. However, it comes at a cost to the environment because the huge heat and chemical changes that are essential in making cement also result in carbon emissions. Cement production is a carbon-intensive process from the viewpoint of the requirement of extraction of enormous amounts of natural resources (raw meal), burning fossil fuels at high temperatures (kiln), as well as process emission of carbon dioxide (CO_2) (Equation 1.1) during the calcination reaction. The total carbon emission arises from cement production through quarrying and preparing raw materials, calcination in kilns for clinker production, milling clinker, and transportation of the final product (cement). Based on the current version of the *Inventory of Carbon and Energy* (ICE) database (Jones, 2019), the embodied carbon content of cement (CEM I) is 0.912 kg CO_2/kg cement, of which 60% of CO_2 (0.55 kg) is released during the calcining reaction, which is necessary for clinker production and cannot be avoided, and 30% emission arises from fuel combustion. A more detailed discussion on approaches to reducing the embodied carbon of cement at the manufacturing level is presented in Chapter 5.

Concrete sector approach to reducing carbon emissions

Globally in 2020, around 14 billion m^3 of concrete and over 4 billion tonnes of cement were produced (GCCA, 2021) to revolutionise the world's built environment. Up to 90% of the carbon emissions associated with the production of concrete are blamed on cement. In 2018, 7.3 million tonnes of CO_2 were emitted by the UK cement industry (MPA UK Concrete, 2020). MPA Cement, which is part of the Mineral Products Association (MPA), represents the six major UK

Portland cement manufacturers that adopted the concrete and cement industry roadmap to deliver net zero emissions by 2050 (MPA UK Concrete, 2020). This roadmap targeted a 39% carbon reduction per tonne of cementitious materials, of which 16% and 12% savings were aimed at the deployment of fuel switching (biomass waste, hydrogen fuel) and low carbon cement (alternative binder), respectively. Likewise, the Institution of Civil Engineers (ICE) in association with the Green Construction Board published the *Low Carbon Concrete Routemap* to roll out and transition to net zero concrete by 2050, recommending three decarbonisation routes, namely benchmarking carbon in concrete, improvement in using concrete, and improvement in making concrete (Low Carbon Concrete Group, 2022). The GCCA, which is a trade association for the cement and concrete sector across the world, pledged contributions in the following items as a pathway to reach the net zero target (GCCA, 2023).

- efficiency in design and construction
- efficiency in concrete production
- savings in cement and binders
- savings in clinker production
- carbon capture, utilisation, and storage at the cement plant
- CO_2 sink/recarbonation
- decarbonisation of electricity.

As a first step, standardised measuring, reporting, and benchmarking of the embodied carbon in concrete for different types of cement need to be established across the construction sector. In 2024 GCCA published global numerical definitions for low carbon and near zero emissions concrete (GCCA, 2024b) for procurement, reporting, and comparing progress between countries, as shown in Figure 1.3, where the band values in global warming potential (GWP) units (ECO_2e/m^3) are plotted against concrete compressive strength (MPa). The global definition of concrete products

Figure 1.3 GCCA global definition of low carbon and near zero emissions concrete. Reproduced with permission from Global Cement and Concrete Association (GCCA, 2024b)

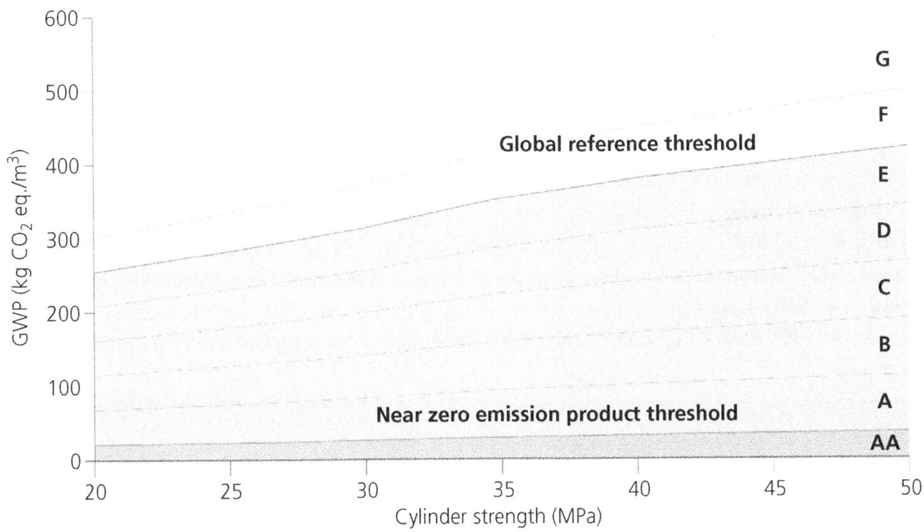

(precast and ready-mix) comprises seven bands (near zero and A to F), including the near zero emissions and global reference thresholds.

To establish the global reference threshold for concrete, data from ten major cement-producing countries (except China) were used to determine carbon footprint thresholds for each country. To create a global threshold, country-specific thresholds were then combined and weighted by each country's cement output.

The International Energy Agency (IEA) proposed 125 kg of CO_2 equivalent per tonne (kgCO2e/t) of cement as the threshold for near zero emissions cement production using 100% clinker, compared to an IEA reference value of 850 kgCO2e/t for a conventional dry kiln (IEA, 2022).

Five equal bands (A to E) are defined after establishing the global reference threshold (top of E) and the near zero emissions threshold (band AA), as shown in Figure 1.3, for the global definition of low carbon and near zero emissions concrete.

The following assumptions were made by GCCA to estimate the near zero emission product threshold for concrete by 2050.

- clinker content in cement will be reduced to 0.52
- cement's carbon footprint will align with IEA's near zero definition
- cement usage per unit volume of concrete will decrease by 14% through advancements in admixtures
- shift to 56-day compressive strength and performance testing.

The unit of carbon intensity or carbon coefficient of concrete products is reported as kg CO_2e/kg, kg CO_2e/t, kg CO_2e/m^2, or kg CO_2e/m^3 by the different stakeholders such as suppliers and developers in the supply chain. Considering this, the ICE *Low Carbon Concrete Routemap* recommended using CO_2e/kg during calculation and CO_2e/m^3 for general reporting and comparison with concrete carbon benchmarks by taking concrete strength class into account (Low Carbon Concrete Group, 2022).

Questions to practice

1. List four major chemical compounds and their corresponding chemical elements commonly found in Portland clinker.
2. Briefly explain how Portland cement is manufactured using natural limestone.
3. Designate a blast furnace cement, conforming to BS EN 197-1 (BSI, 2019a), containing between 81% and 95% by mass granulated blast furnace slag (S) of strength class 32.5 with low early strength.
4. List a few strategies that can be adopted to reduce carbon emissions in the concrete sector.

Monower Sadique
ISBN 978-1-83608-477-8
https://doi.org/10.1108/978-1-83608-474-720251002

Chapter 2
Concrete

After completing this chapter, you should be able to

- understand the fundamental properties of concrete
- identify the basic constituents of concrete and their roles
- explain the principles of concrete mix design
- recognise the relevant standards governing concrete specifications
- distinguish between different classes of concrete and their applications
- describe how laboratory trials for concrete are conducted and how repeatability is ensured.

Introduction

Concrete is a composite material made of coarse aggregate, fine aggregate, cement or binder, and water, with each constituent playing a crucial role in the fresh and hardened properties of concrete. Careful selection and proportioning of these constituents, typically achieved through mix design, are essential to achieve the desired performance for fresh and hardened concrete, including workability, density, strength, and durability. The choice of materials affects the density of the concrete, while its strength at maturity determines the strength class used for classification. To ensure consistent concrete strength with high confidence, the statistical significance and repeatability of the mix design should be validated through multiple trials.

Concrete, mix design, and standards

According to BS EN 206:2013+A2:2021 (BSI, 2021c), concrete is a monolithic material formed by mixing cementitious binder, fine and coarse aggregate, and water, with or without the incorporation of admixtures, additions, or fibres, which develops its properties by hydration reaction. In 2020, 14 billion tonnes of concrete was consumed globally (GCCA, 2021), whereas the UK's annual consumption is 90 million tonnes (MPA UK Concrete, 2020). Concrete's resistance to fire and abrasion, thermal conductivity, specific heat capacity, and strength have attracted its widespread application in the construction sector. Concrete is a brittle material with an elastic modulus in compression that varies between 14×10^3 and 4×10^3 MPa, while its thermal conductivity value ranges from 1.5±3.5 W/m/K for normal weight concrete to 0.5–1.1 W/m/K for lightweight concrete.

The cementitious binder in concrete can be Portland clinker only or fine powder additions (GGBS, PFA, SF, etc.) that partially replace clinker and participate in the hydration reaction. The admixtures are chemicals that can be added to concrete at not more than 5% (of binder) to alter, modify, or improve the fresh and hardened properties of concrete. Aggregates with a particle size of less than 4 mm are termed fine aggregates and all larger aggregates with a maximum diameter of 40 mm are termed coarse aggregates. The binder which is the key constituent of concrete occupies the smallest volume (*c*. 10%), whereas combined aggregates (coarse and fine) occupy the largest volume (*c*. 80%) in concrete.

To meet workability (placement, compaction cohesiveness, mobility, and surface finish), strength (compressive), and durability (minimum binder content, maximum water/binder ratio, etc.) requirements, appropriately mixed concrete is essential. The process for the selection of content or proportion of each constituent of concrete is termed a mix design. Concrete mix design is an important process that defines the proportion of each constituent to be mixed to achieve the desired strength and properties. Concrete mix design is largely dependent on the constituent materials' properties, such as cement type, aggregate type (crushed or uncrushed), relative density and their grading (particle size distribution (PSD)), desired workability and so on. In the UK, the Building Research Establishment (BRE) provides detailed procedures for mix design (BRE, 1997) using empirical formulae and relationships, with worked examples incorporating various scenarios.

The concrete mix design specifies the weight of each constituent material per unit volume of concrete (kg/m^3) – for example, following the BRE's prescribed procedure (BRE, 1997) for the mix design of given concrete will provide the dosage of each constituent with concrete density as follows

total cementitious binder: 300 kg/m^3
fine aggregate: 600 kg/m^3
coarse aggregate: 1200 kg/m^3
free water: 150 kg/m^3

---

concrete density: 2250 kg/m^3

Typically, the density of concrete using normal weight aggregates ranges from 2200 kg/m^3 to 2400 kg/m^3. The final mix proportion can also be expressed in terms of ratio relative to each part of the binder using a binder/fine aggregate/coarse aggregate ratio as 1:2:4, either by mass or volume, or as binder/aggregate ratio, water/binder ratio and fine aggregate/coarse aggregate ratio, usually by mass. However, given the complexities in discussing the influence of mix parameters on the concrete and required information in terms of the quantity of constituent materials necessary to cast a given volume of concrete, the BRE guidance document (BRE, 1997) suggests mix parameters as the mass (kg/m^3) of each constituent material needed to produce one cubic metre of concrete.

The cementitious binder in the final mix design can be with or without any additions. If additional materials are used, their proportions must be adjusted based on their replacement level relative to the total binder content – for example, if Portland silica fume cement with 6% silica fume content is proposed, calculations (silica fume (6% of binder = 18 kg/m^3); Portland cement (94% of binder = 282 kg/m^3) show that the mix design would be adjusted to include 282 kg/m^3 of Portland cement and 18 kg/m^3 of silica fume, maintaining a total binder content of 300 kg/m^3.

It is also essential to note that water added to concrete is absorbed by the aggregate and is also needed for both the hydration reaction and workability of fresh concrete. Therefore, to ensure consistent concrete strength and workability, using the term 'free water/binder ratio' rather than 'water/binder ratio', with aggregates considered in a saturated surface dry (SSD) condition, is a more accurate approach in concrete mix design.

BS EN 206:2013+A2:2021 (BSI, 2021c), in conjunction with the complementary standards BS 8500-1 (BSI, 2021a) and BS 8500-2 (BSI, 2021b), gives national provisions that prescribe the method of specifying concrete in the UK. The different properties of fresh concrete (such as workability, air content, etc.) are controlled by 12 parts of the BS EN 12350 (BSI, 2019b) series, whereas concrete hardened properties (such as density, strength, etc.) are controlled by 18 parts of the BS EN 12390 (BSI, 2019c) series.

Concrete class

BS EN 206:2013+A2:2021 (BSI, 2021c) classifies concrete based on fresh and hardened properties. At fresh state, normal concrete classes are slump (S1 to S5), compaction (C0 to C4), and flow (F1 to F6). For self-compacting concrete (SCC), classes are viscosity (VS1, VS2, VF1, VF2) and passability (PL1, PL2, PJ1, PJ2). All of these are based on the consistency of the concrete. In the case of hardened concrete, the classes are based on compressive strength for normal and heavyweight concrete, and density (D1.0 to D2.0) for lightweight concrete. When testing concrete in the laboratory, samples can be in the form of cubes with a nominal edge length of 150 mm or cylinders with a nominal diameter of 150 mm and a length of 300 mm. Concrete strength classification includes the type of concrete (normal weight, heavyweight, or lightweight), the minimum characteristic cylinder strength, and the cube strength. This classification is denoted by the letter 'C' for normal weight and heavyweight concrete, or 'LC' for lightweight concrete, followed by two numbers representing the characteristic compressive strength at 28 days for cube and cylindrical specimens, respectively, separated by a forward slash. According to BS EN 206:2013+A2:2021 (BSI, 2021c), the strength class for normal weight and heavyweight concrete ranges from C8/10 to C100/115, while for lightweight concrete, the strength class ranges from LC8/9 to LC80/88.

Laboratory trials

After the initial calculations during the mix design, the design needs to be verified for desired strength, workability, and other properties through trial mixes on a small scale in the laboratory. Based on the results obtained from the first trials, if the results (strength, workability) do not meet the desired specifications, further adjustments to the mix proportions should be made.

For instance, if initial trials reveal adequate strength but insufficient workability (e.g. low flow or slump), adjustments can be made in subsequent trials, such as increasing the water/binder ratio or adding a plasticiser while maintaining the same water/binder ratio. This iterative process continues until both fresh and hardened concrete properties meet target values with satisfactory repeatability. However, consistent results in the lab do not guarantee identical in situ properties, as variations in mixer type, concrete volume, placement, compaction, aggregate moisture content and so on can affect fresh and final properties. Therefore, regular testing of representative in situ samples (based on concrete quantity) under a quality control programme is essential to confirm consistency.

Due to the inherent variabilities of the constituent materials, targeting a considerably higher mean strength than the strength specified during the designing of a concrete mix will leverage the risk. Considering variations in concrete strength follows the normal distribution as shown in Figure 2.1, the target mean strength can be estimated as

Figure 2.1 Typical normal distribution of concrete strength (author's own)

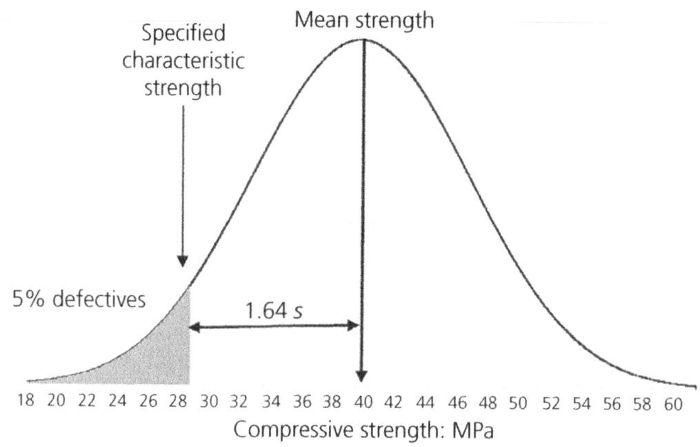

$$f_m = f_c + ks \quad (2.1)$$

where f_m is the target mean strength
f_c is the specified characteristic strength
s is the standard deviation (SD)
k is a constant.

To give 95% confidence (5% defectives) in achieving the characteristic strength, the value of $k = 1.64$ with a total margin of 1.64 x SD is usually suggested. Typically, the standard deviation of the strength of standard concrete cubes at 28 days varies between 3 N/mm^2 and 5 N/mm^2. The BRE design guides (BRE, 1997) recommend using the standard deviation calculated from at least 20 samples. If fewer samples are available, a default standard deviation of 8 N/mm^2 is suggested.

Specimen preparation and loading are crucial for testing concrete strength, as governed by BS EN 12390-3 (BSI, 2019c). The standard outlines acceptable and unacceptable failure modes of concrete samples, contingent on proper sample positioning and loading rates. In cases of unsatisfactory failure, the standard recommends recording the failure type in the test report concerning the mode classified in the standard; these failures may result from various influencing factors. Illustrations of acceptable failure modes for cube and cylinder samples are shown in Figure 2.2.

Figure 2.2 Satisfactory failure of concrete: (a) cube specimen and (b) cylinder specimen (BSI, 2019c). Courtesy of BSI

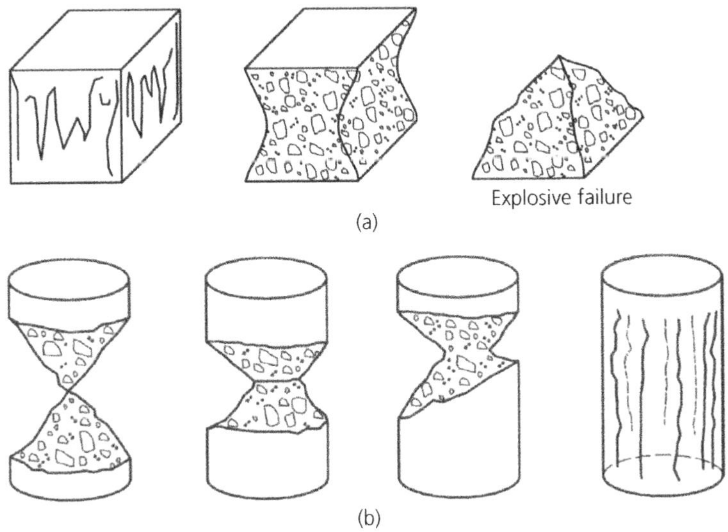

Monower Sadique
ISBN 978-1-83608-477-8
https://doi.org/10.1108/978-1-83608-474-720251003

Chapter 3
Concrete strength and hydration reaction

After completing this chapter, you should be able to

- understand how concrete develops strength through the hydration process
- explain the hydration reaction, its stages, and the progression of the reaction over time
- identify the hydration products that impart strength to concrete and their roles
- analyse how the progression of hydration can be observed and measured using analytical tools
- explore the factors influencing hydration reactions and their impact on concrete properties
- comprehend the role of key concrete constituents in determining fresh and hardened properties
- examine different types of aggregates used in concrete and understand how their gradation affects concrete performance
- understand the influence of the water/binder ratio on the workability, strength, and durability of concrete
- define concrete workability and identify the factors that control it
- understand the causes and consequences of segregation and bleeding in concrete
- define admixtures and describe their purposes in concrete applications
- explore the types of admixtures commonly used in concrete and their specific roles
- evaluate how plasticisers affect the workability and strength of concrete
- determine when set-accelerating and set-retarding admixtures are appropriate for use
- explain the role of air-entraining admixtures in enhancing concrete's resistance to freeze–thaw cycles
- explore how admixtures contribute to the sustainability of concrete production and usage
- understand the specifications, advantages, and applications of precast concrete
- explore the properties and applications of roller-compacted concrete in various construction contexts.

Introduction

The strength of concrete develops through the hydration reaction, a multi-phase process that transforms a hydraulic cementitious binder into a concrete-like ceramic material. This complex reaction is influenced by various external and constituent factors, impacting both the fresh and hardened properties of concrete. Additionally, beyond the hydration reaction, concrete's characteristics, durability, and performance are shaped by its components, all of which must be carefully considered when designing a mix for a specific application. The primary objective in concrete design is to create a monolithic, high-density structure with minimal porosity, preventing the ingress of moisture and harmful gases during service. Achieving this requires careful control of mixing, transporting, placing, compacting, and curing processes, all of which are critical to the concrete's durability and performance.

Hydration reaction

In chemistry, the term hydration refers to the reaction of an anhydrous material with water resulting in a new product called hydrate. In cement chemistry, hydration is defined as the exothermic reaction of anhydrous cement with water; the resulting setting and hardening and hydration product or hydrate is a new compound formed through hydration. Cement hydration is an instantaneous and spontaneous reaction where fresh cement paste is converted to hardened cement paste (HCP) as hydration progresses. Hydration is a sequence of complex chemical reactions that converts the anhydrous phases of cementitious material into hydrated phases and liberates heat, as shown in Equation 3.1, and paste (pure cement with water at a water/cement ratio of 0.4), as shown in Figure 3.1. In the aqueous environment where hydration progresses, water is bound both chemically (to form hydrates) and physically (by absorption into pores and adsorption onto the surface of the hydrates).

$$\text{Anhydrous binder} + \text{water} \rightarrow \text{hydrates} + \text{heat} \uparrow \qquad (3.1)$$

Figure 3.1 Typical paste of CEM I at 0.4 water/cement ratio (author's own)

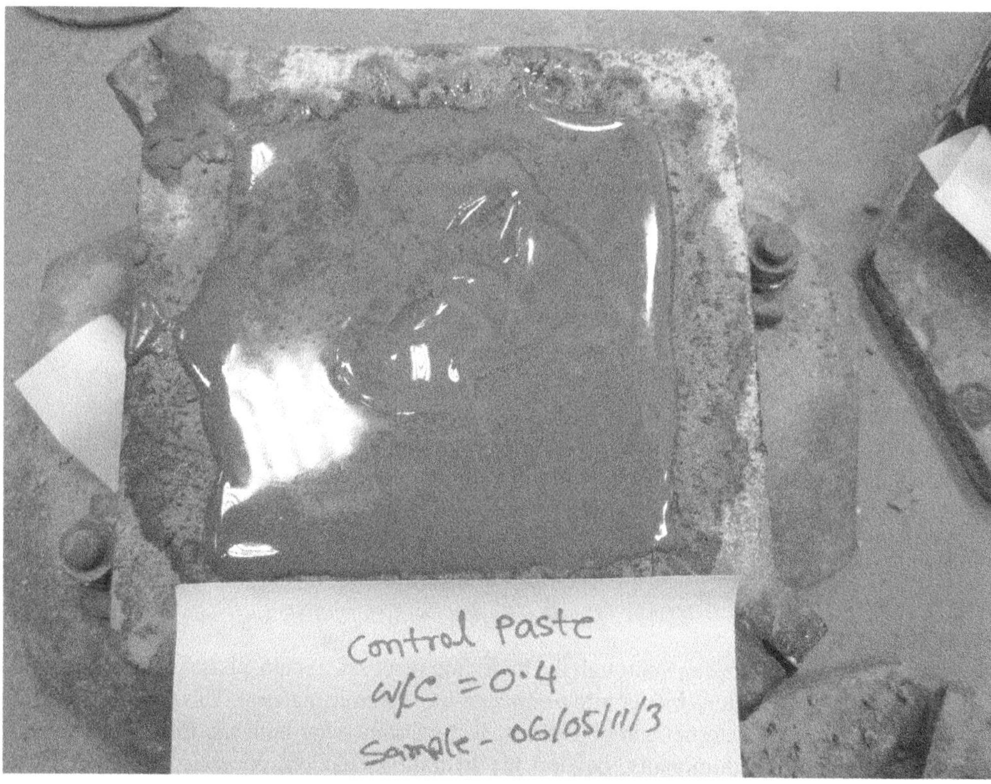

Stages of hydration reaction

Beaudoin and Odler (2019) and Soutsos and Domone (2018) describe the hydration process of cement with the following four stages; however, due to the multi-component system of Portland clinker, the chemical reactions can occur either in parallel and successively (Lea *et al.*, 2019). The products (hydrates), namely calcium silicate hydrate or C-S-H gel, portlandite or $Ca(OH)_2$ or C-H, ettringite (AF_t phases), monosulfate (AF_m phases) and hydrogarnet phases have been suggested by Gollop and Taylor (1994).

(*a*) Pre-induction period (first minutes). Immediately after the interaction of cement with water, an ionic species alkali sulfate (K^+, Na^+, and SO_4^{2-}) dissolves in the liquid phase and forms hydrates. At the same time, $CaSO_4$ dissolves and contributes Ca^{2+} and additional SO_4^{2-} ions until saturation. Tricalcium aluminate (C_3A) of cement dissolves at this stage and reacts with Ca^{2+} and SO_4^{2-} and precipitates as ettringite ($C_3A{\cdot}3CaSO_4{\cdot}32H_2O$) or AF_t phase.

(*b*) Dormant period (first few hours). For a period of a few hours the overall reaction slows down, and the cement is relatively inactive. The concentration of Ca(OH) reaches its maximum and the concentration of SO_4^{2-} ions remains constant as the fraction consumed during ettringite formation is replaced by dissolution of additional $CaSO_4$. The nucleation of the calcium silicate hydrate (C-S-H) phase initiates at this stage and the initial set of cement also occurs at this stage without gaining any strength or hardening.

(*c*) Acceleration stage (3 to 12 hours). The hydration of tri-calcium silicate (C_3S), as well as dicalcium silicate (C_2S), accelerates at this stage and C-S-H starts to be formed. The concentration of Ca^{2+} and SO_4^{2-} ions in the liquid phase gradually declines due to the formation of portlandite (C-H) and ettringite, respectively. The adsorption of SO_4^{2-} ions onto the surface of C-S-H also reduces the ion concentration of sulfates.

(*d*) Post-acceleration period. The C-S-H phase continues to be formed for continuing hydration of C_3S and C_2S. At the same time, due to a shortage of SO_4^{2-} ions as a consequence of the complete dissolution of $CaSO_4$, the previously formed ettringite (AF_t) phase is converted to monosulfates ($C_3A{\cdot}CaSO_4{\cdot}12H_2O$) or AF_m.

The entire process is presented in Figure 3.2.

Understanding and analysis of hydration

The scanning electron microscope (SEM) imaging technique has been used widely in the cement research sector for petrographic analysis of hardened products and to estimate the degree of hydration as it provides elemental composition and helps to discriminate between different phases, as shown in Figure 3.3. SEM is a method for high-resolution imaging of surfaces. C-S-H has been termed the most important component in concrete since it provides binding property (through Van der Waals forces) and contributes strength. At the same time, SEM was found to be exceptionally useful for characterising the morphology and chemical composition of the C-S-H phase (Sarkar *et al.*, 2001). The morphology of the different phases of the hydrated cement paste can be summarised as follows, based on a study by Sarkar *et al.* (2001).

- The C-S-H phase in cement paste produced at normal temperature is represented by gel structure.
- The C-H crystals appear in many different shapes and sizes, starting from massive, platy crystals with distinctive hexagonal prism morphology or large, thin, elongated crystals.
- The ettringite (AF_t) forms needle-like crystals in vacant spaces during early hydration.
- Monosulfate (AF_m) appears as hexagonal, platy crystals.

Figure 3.2 Hydration kinetics of CEMENT (Beaudoin and Odler, 2019)

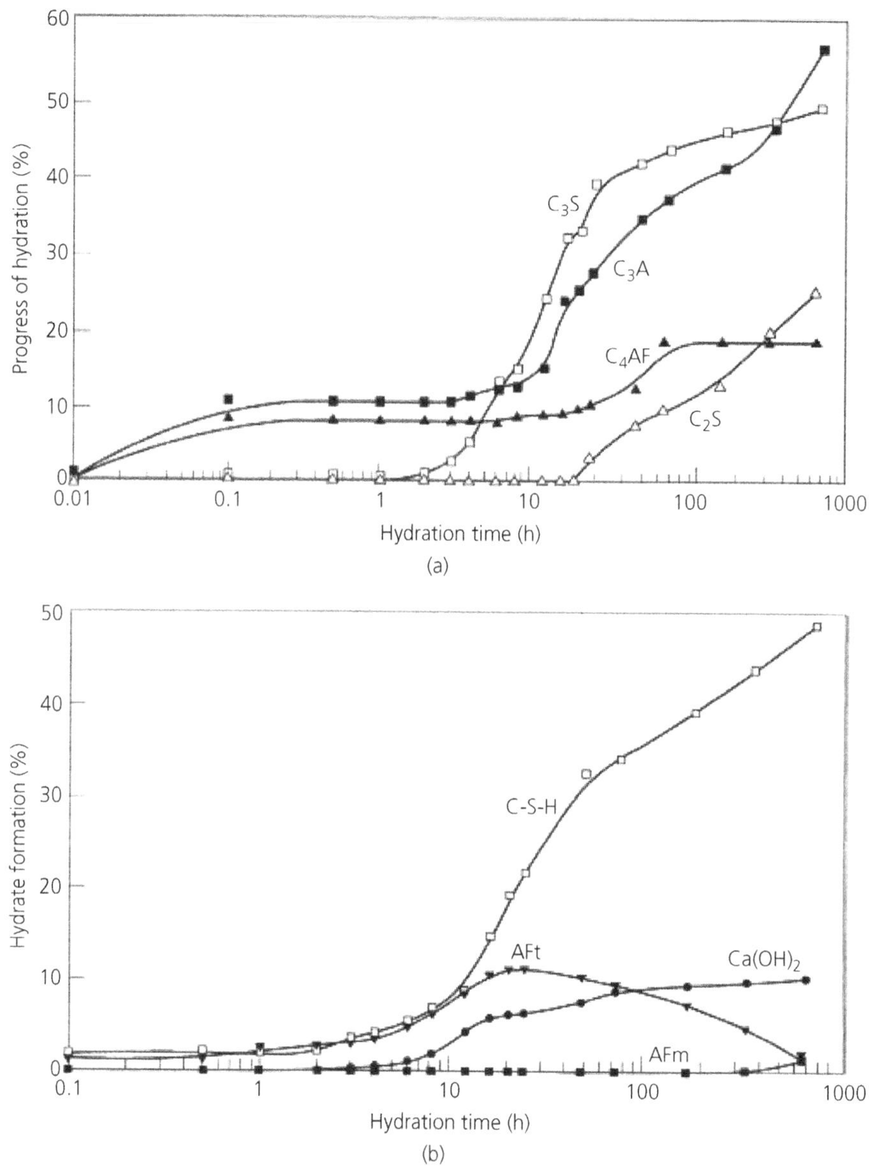

Likewise, the following general features of reaction sequence and hydrates of cement concerning X-ray diffraction (XRD) analysis have been stated by Chatterjee (2001).

- Unreacted phases in different proportions coexist along with various reactants at different ages.
- The silicate phases lead to the formation of C-H and C-S-H, the latter being amorphous or poorly crystalline.

Figure 3.3 SEM micrograph of hydrated cement pastes at different ages: (a) needles of ettringites, (b) crystals of portlandite (C-H) and (c) C-S-H gel (author's own)

(a)

(b)

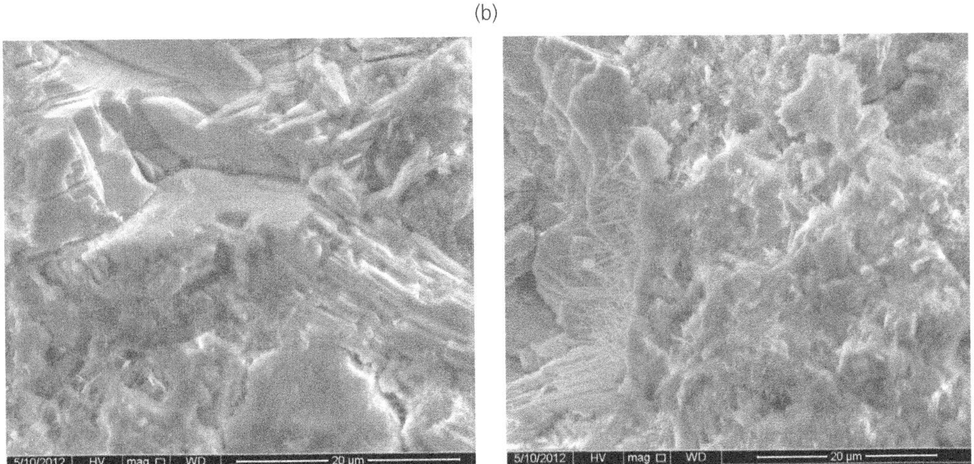

(c)

Figure 3.4 Progress of hydration of cement in XRD (Sadique *et al.*, 2013), Courtesy of Elsevier

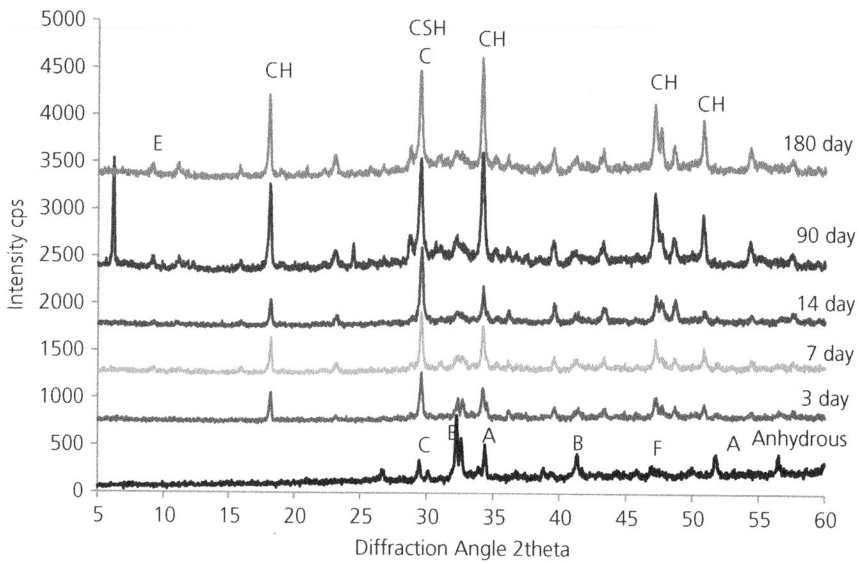

- The reactions of the aluminate and ferrite phases in the presence of gypsum lead to the early formation of AF_t. The most important AF_t phase is ettringite $[(Ca6Al_2(SO_4)_3(OH)_{12}\cdot 26H_2O]$, which forms during the early hydration of cement and later converts to AF_m $(C_3A\cdot CaSO_4\cdot 12H_2O)$.

Moreover, the utilisation of XRD for tracking the hydration reaction over time, which is accompanied by the consumption of mineral phases and their subsequent reduction of crystal nature, was reported by Esteves (2011). Typical XRD results for the hydrated cement paste are displayed in Figure 3.4 (Sadique *et al.*, 2013).

Interfacial transition zone (ITZ)

Concrete is a three-phase material comprising HCP, aggregate, and the ITZ, a thin region of cement/binder pastes around the aggregates that govern bonding (chemical and mechanical) and strength at the microscopic level. The density of ITZ is considerably lower than the aggregate and different from the rest of the paste. The performance characteristics of ITZ (denser, rich in C-S-H hydrates, and low in C-H and ettringite) make concrete more impermeable and durable, with higher strength, particularly when the water/cement ratio is minimised, although achieving completely impermeable concrete is practically impossible. A denser (Shi and Xie, 1998) and more homogeneous (San Nicolas *et al.*, 2014) ITZ in alkali-activated slag compared to cement has been reported, whereas in the case of alkali-activated fly ash, gel-rich paste with narrower and denser microstructure in ITZ with less portlandite (C-H) compared to cement has been reported (Luo *et al.*, 2021).

Ongoing hydration increases C-S-H content, reducing porosity and enhancing ITZ density. In normal strength concrete, however, the ITZ is often the weakest link, where micro-cracks can develop under structural and environmental loads and subsequently propagate through concrete leading to failure. Factors like aggregate size and grading, aggregate mineralogy and surface texture, cement content, water/cement ratio, and curing conditions influence the amount of micro-cracking in the ITZ and, consequently, the overall concrete performance. Notably, larger aggregate sizes create a higher local water/cement ratio in the ITZ, resulting in weaker, more permeable concrete (Mehta and Monteiro, 2014). The effect of aggregate size and water/cement ratio on the thickness of ITZ and its impact on concrete mechanical properties was researched by Wang *et al.* (2024a). In this research, the thickness of the ITZ was found to inversely impact the bonding force between the coarse aggregate and cement matrix, leading to reduced compressive strength and elastic modulus. This effect was attributed to the effective transfer of load and decreased stress concentration in the ITZ as its thickness diminished. Additionally, a larger coarse aggregate was reported to induce a thicker ITZ with a water/cement ratio of 0.3, whereas the opposite correlation was found for a water/cement ratio of 0.4. Likewise, owing to the lower overall surface area of larger aggregates, a weaker transition zone, with a more critical effect on the concrete's strength at a lower water/cement ratio, was suggested by Soutsos and Domone (2017).

Factors affecting hydration

The rate of progress of hydration reaction and associated strength gain by a cementitious binder are influenced by the following key factors.

- The distribution of different phases/compounds. The hydration reaction is largely dependent on the distribution of phases or compounds in cement. Cement with a higher proportion of C_3A and C_3S will set more quickly, whereas an increased proportion of C_3S and decreased C_2S content will result in rapid strength gain but is associated with liberation of higher heat.
- The fineness (specific surface area, SSA) of cement/binder. The fineness of cement or cementitious binders influences their reactivity and affects the rate of hydration. A cement with finer particles with higher SSA will participate in the hydration reaction quicker than its coarser counterparts and influences the setting and hardening of concrete and mortar.
- Used water/binder ratio. A minimum water/cement ratio of approximately 0.43 is required to fully hydrate all cement particles, while a water/cement ratio below about 0.38 results in incomplete hydration (Soutsos *et al.*, 2017).
- Presence of admixtures and additives. The presence of any chemical admixture such as an accelerator or retarder added to modify the setting time of concrete will influence the rate of hydration.
- Curing regime. The main purpose of curing is to control and ensure enough supply of moisture during hydration. In the initial stage, the rate of hydration of Portland cement increases with higher curing temperatures. However, as hydration progresses, the influence of temperature becomes less pronounced. At lower temperatures, such as concreting in winter when the temperature reaches –10°C, the rate of hydration is very low, and concreting becomes very problematic. At the same time, the condition of curing (such as water- or air-cured) also significantly affects rate and degree of hydration. As shown in Figure 3.5, after 180 days, an air-cured concrete was reported to achieve only 40% of the strength compared to its water-cured counterpart. Curing can done by spraying, steaming, ponding, covering with waterproof sheets, or applying a curing membrane.
- Age (of curing). Although strength development in concrete depends on many factors, it is predominantly influenced by the chemical hydration reaction. The rate of hydration gradually increases with time, leading to a corresponding increase in paste strength, as shown in Figure 3.6. Typically, the degree of hydration completes approximately 80%

Figure 3.5 The effect of curing regime on concrete strength (contains example data only). Reproduced with permission from Cement Concrete & Aggregates Australia (CCAA, 2006, Part V, s.15, p. 4, figure 15.3)

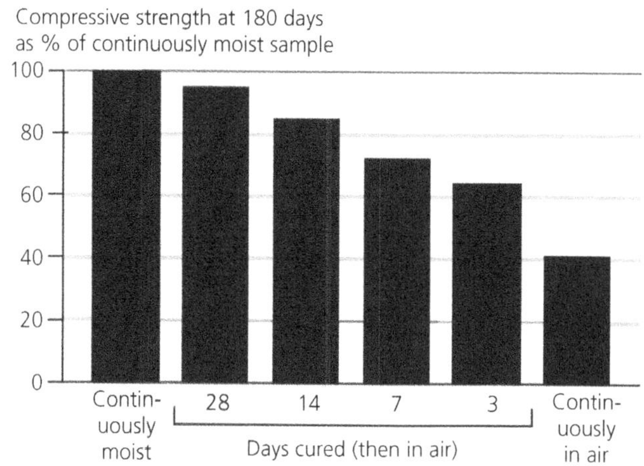

Figure 3.6 Influence of hydration and age on strength development of Portland cement paste (Singh, 2023)

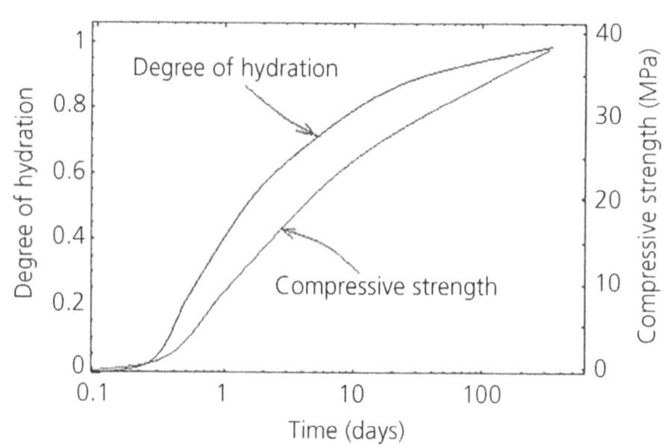

at 28 days, as shown in Figure 3.6, and for this reason, the 28-day strength is commonly specified as an indicator of final strength in the construction sector and in cement and concrete standards for quality control purposes, although the hydration reaction may take several years to fully complete.

Key concrete properties

Concrete possesses several inherent versatile properties, such as being non-toxic and inert, having high thermal mass, mouldability in its fresh state, fire resistance, and high density. These properties, resulting from the constituent materials used, contribute to its widespread application in the construction sector. In the UK, the key properties of fresh concrete, such as flow or consistency, density, and compatibility, are controlled by different series of BS EN 12350 (BSI, 2019b). The key hardened properties, such as compressive, tensile, and flexural strength, permeability, and density, are governed by various series of BS EN 12390 (BSI, 2019c). The properties of concrete, both in its fresh and hardened states, are significantly influenced by the constituent materials, which are determined through the structural design process followed by mix design. Besides the careful selection of these materials, the development and use of modern admixtures can modify and achieve the desired concrete properties, thereby contributing to the sustainability of concrete structures. The selection of concrete proportions involves a balance between economy and requirements of workability, strength, durability, density, and appearance. The flexibility between these multi-dimensional characteristics of conventional cement makes it very versatile. Some key concrete constituents and how they influence concrete properties are briefly discussed below.

Aggregates

As mentioned earlier, aggregates (fine and coarse) which are an inert filler comprise around 60% to 80% of the concrete volume, and their properties significantly influence the characteristics of concrete in both its fresh and hardened states. Factors such as aggregate type (density), gradation (particle size distribution), proportion (aggregate/binder ratio), shape, and surface properties (roughness, crushed or uncrushed) play a crucial role in determining the overall performance of the concrete. Aggregates used in concrete can be natural, artificial, reclaimed, or recycled granular minerals. BS EN 12620 (BSI, 2002) specifies the properties of aggregates in concrete. Fine aggregates are defined as those with sizes less than or equal to 4 mm, while aggregates with sizes greater than or equal to 4 mm are termed coarse aggregates. Natural aggregates either crushed or uncrushed are sourced from quarry igneous, sedimentary, or metamorphic rocks and processed to desired size and grading. Artificial or manufactured aggregates are sourced either from other industrial processes (such as blast furnace slag, waste glass, etc.) or specially manufactured (such as expanded clay or sintered fly ash).

The oven-dry particle density of aggregates also determines the type of concrete: heavy, light, or normal weight. Concrete made with aggregates of mineral origin having an oven-dry particle density $\leq 2000\,kg/m^3$ (expanded shale, foamed blast furnace slag, etc.) is termed lightweight concrete, aggregates with a particle density of $2000–3000\,kg/m^3$ (basalt, limestone, sandstone, etc.) create normal weight concrete, and those with a density $\geq 3000\,kg/m^3$ (haematite, manufactured steel shot, etc.) create heavyweight concrete.

Different physical and durability properties of coarse aggregate such as resistance to fragmentation, resistance to wear, resistance to polishing, resistance to abrasion, water absorption, freeze–thaw resistance, and alkali–silica reactivity need to be measured as specified in BS EN 12620 (BSI, 2002). Concrete performance against alkali–silica reaction, fire, and freeze–thaw resistance significantly depend on aggregates. In the case of prolonged exposure to fire, the differential volume of concrete changes due to the contraction of HCP, and expansion of aggregates leads to thermal stress followed by cracks. The breaking up of Thames gravel, limestone aggregate, and carbonate aggregate due to differential expansion at 350°C, 650°C, and above 700°C was reported by Khoury (1992). The thermal stability of aggregates, both chemically and physically, needs to be considered for enhanced fire resistance of concrete.

Aggregate obtained after processing hardened concrete that has not been previously used in construction is termed reclaimed aggregate, whereas aggregate recovered by processing inorganic material previously used in construction is termed recycled aggregate. Reclaimed aggregate can be obtained by washing fresh concrete or crushing unused hardened concrete.

BS EN 206 (BSI, 2021c) specifies the quantities of undivided (coarse and fine) reclaimed aggregate shall not be greater than 5% by mass of the total aggregate. In the case of quantities of reclaimed crushed aggregates greater than 5% they shall be treated as recycled aggregates.

The ICE database (Jones, 2019) lists aggregates contributing 3.9% of the total embodied carbon (377 kg CO_2e/m^3) in a CEM I-based precast concrete mix, highlighting the importance of using recycled aggregates to reduce environmental impact. Construction and demolition waste (C&DW) is a significant global challenge and recycling this waste into recycled aggregate (RA) for new concrete is becoming a common practice to improve circularity in the construction industry. There are two main types of recycled aggregates derived from C&DW.

- mixed recycled aggregates (MRAs): these are heterogeneous, produced in larger quantities, and are typically unsuitable for use in structural concrete
- recycled concrete aggregates (RCA): these have lower heterogeneity and are better suited for structural concrete production.

BS EN 206 (BSI, 2021c) recommends replacing 30–50% of natural coarse aggregates with coarse recycled aggregates, depending on the exposure class, for structural concrete. The *Design Manual for Roads and Bridges* (CD 374) (Highways England, 2020) allows a maximum of 60% recycled aggregate by mass in non-prestressed concrete and 20% in prestressed concrete structures. This supports the use of recycled materials while ensuring the structural integrity of concrete. Although the mechanical strength of traditional concrete is often negatively impacted by increasing the content of RA, only a slight reduction in compressive and splitting tensile strength was observed in geopolymer concrete with higher RA content (Nuaklong *et al.*, 2016). This reduction was further mitigated by adjusting the sodium silicate dosage. A decrease in strength and an increase in water absorption were reported in geopolymer mortar when natural sand was replaced with recycled brick dust (Roy and Sadiqul Islam, 2024). Similarly, the use of recycled bricks as coarse aggregate in geopolymer concrete yielded unsatisfactory results, indicating challenges in maintaining performance with such substitutions (Roy and Sadiqul Islam, 2024). The strength development of fly ash- and GGBS-based geopolymer concrete incorporating 100% RA, as investigated by the author, was found to be comparable to that of conventional concrete made with 100% natural aggregate (NA), as illustrated in Figure 3.7. Remarkably, the strength of the geopolymer concrete exceeded that of the control mix utilising 100% NA. A ratio of fine aggregate to coarse aggregate of 1:2.3 was used. A similar particle size distribution between recycled and natural aggregates was used in this investigation, as shown in Figure 3.8.

Grading of aggregates

Concrete is a three-phase composite material consisting of coarse aggregate, fine aggregate, and cement/binder paste. The PSD or grading of coarse and fine aggregates significantly influences their workability, permeability, density, and required binder content for achieving the desired strength. For aggregate particle packing theory, the grading of coarse and fine aggregates individually and combined must be determined by sieve analysis according to BS EN 12620 (BSI, 2002) before mix design and laboratory trials are carried out to determine optimal proportions. Several empirical theories have also been developed to optimise the gradation of concrete with enhanced properties. Concrete's workability, thixotropy, and required binder content for a design strength are significantly influenced by aggregate gradation and shape. Generally, a higher proportion of coarse aggregate (CA) than fine aggregate (FA) generates a coarser grading which lowers the required binder content for the desired strength and consistency. Continuous smooth-graded

Figure 3.7 Strength development of concrete made with natural aggregate (NA) and geopolymer (Geo) concrete made with NA and recycled aggregate (RA) (author's own)

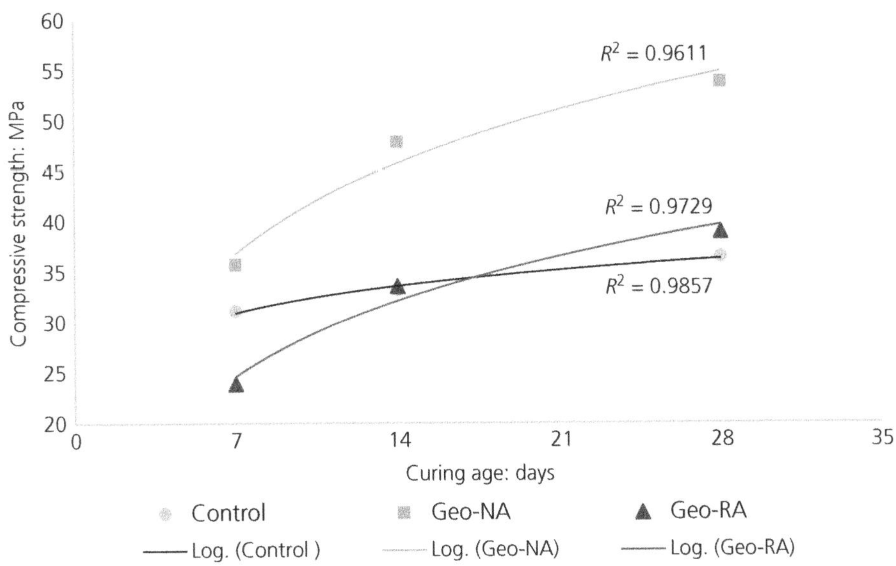

Figure 3.8 Particle size distribution of natural and recycled aggregates used in concrete (author's own)

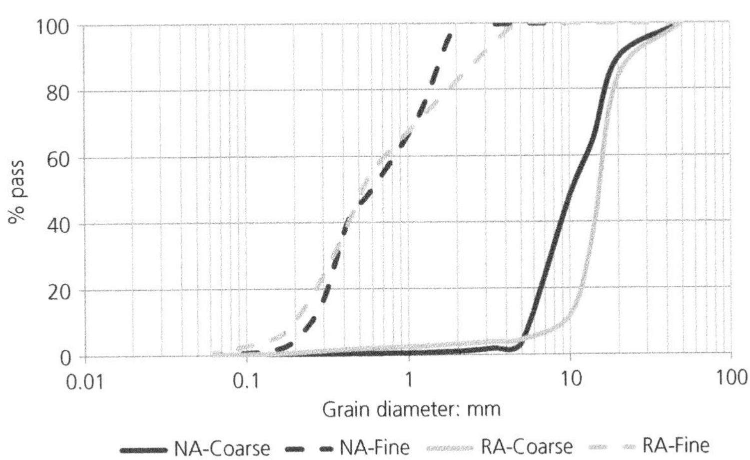

aggregates create an economic mix with fewer voids. Figure 3.9 shows typical individual and combined PSD of aggregates and illustrates how their composition (CA/FA ratio) influences the overall grading. A concrete having $\leq 20\%$ coarse aggregate retained on a given sieve size is reported to show a uniform distribution of coarse aggregate resulting in satisfactory workability with reduced chances of segregation (Sokhansefat *et al.*, 2019). Likewise, for reduced void content and higher workability, a continuous range of particle sizes from fine sand up to coarse aggregates was suggested (Soutsos and Domone, 2018). For a given coarse aggregate content, a

Figure 3.9 The influence of coarse and fine aggregate proportion on combined aggregate gradation (author's own)

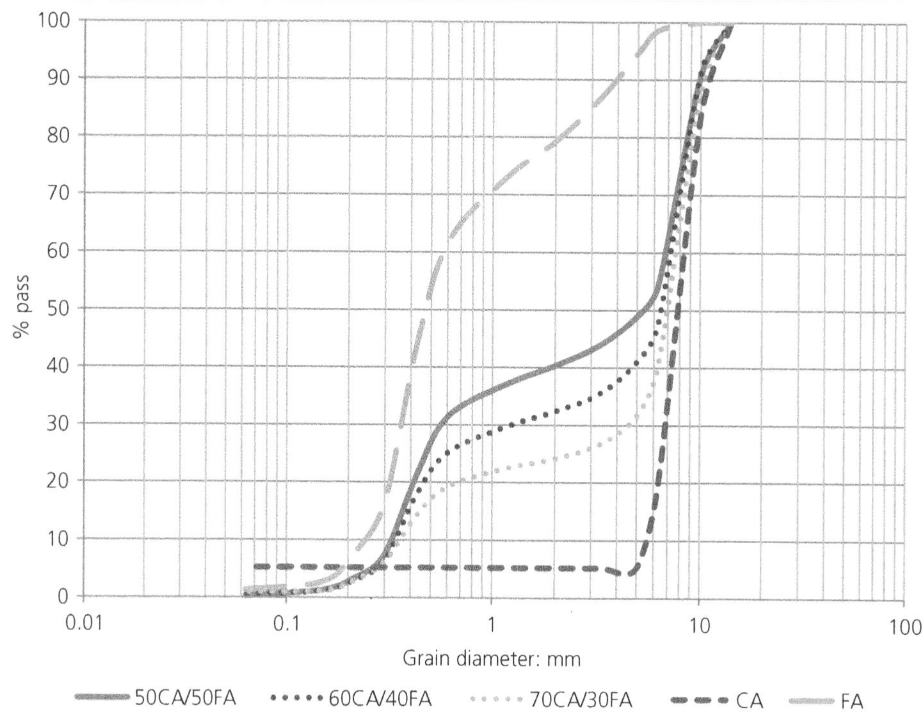

continuous grading is reported to provide the highest compressive strength compared to the gap graded in the case of normal strength concrete (NSC) with the contrary for high-strength concrete (HSC) (Meddah *et al.*, 2010).

The fineness modulus (FM) is a widely accepted empirical measure used to indicate the gradation of aggregate, whether coarse or fine. It is calculated by dividing the sum of the cumulative percentages retained on selected sieves during sieve analysis by 100. A higher FM value indicates a coarser aggregate. Considerable influence on concrete strength and workability for varying FM of fine aggregate compared to that of combined aggregate was reported (Ashraf and Noor, 2011). Additionally, a 50% increase in compressive strength was observed with varying combined aggregate gradation. Moreover, the requirement of less water/cubic metre for mixes with larger maximum aggregate size to achieve a given slump increase was reported (CCAA, 2020).

Water/binder ratio

The water/cement or water/binder ratio is one of the most important parameters that significantly influences concrete or mortar properties in fresh and hardened states. The influence of water/cement ratio on concrete strength was first reported by Abrams (1919), significantly advancing concrete technology. According to Abrams' law, concrete compressive strength (f_c) is inversely proportional to the water/cement ratio which can be expressed by the following empirical relationship

$$f_c = \frac{A}{B^{\frac{W}{C}}} \tag{3.2}$$

where w/c is the water/cement ratio and A and B are empirical constants that depend on given materials, test conditions, and age.

Although the empirical relationship and the inverse curve of strength against water/cement ratio are evident, Abrams also highlighted in his pioneering publication that this relationship is valid as long as the mix is workable. At very high water/cement ratios, the cement paste becomes highly porous, reducing the strength of the concrete. Although workability increases at this stage, this can lead to segregation. On the other hand, at very low water/cement ratios, the workability of concrete drastically decreases, making it difficult to achieve complete compaction. This results in concrete with higher void content and lower strength. To address this, using a suitable water-reducing admixture to increase consistency without increasing the water/cement ratio is a widely accepted practice, thanks to advancements in modern plasticisers and superplasticisers. Based on the experimental results of cement mortar at varying cement/sand ratios, a generalised relationship was developed (Singh *et al.*, 2015b) as a function of the water/cement ratio using regression analysis, as shown in Figure 3.10. The strength against water/cement curve shown in Figure 3.10 could be further moved up and left by using enhanced compaction and advanced superplasticiser.

For HSC or ultra-high-strength concrete (UHSC) reduced porosity is crucial; therefore, a low or very low water/binder ratio is used, increasing the chances of self-desiccation which can be minimised by appropriate curing. In pervious concrete, where binders are primarily used to form a thick coating around the aggregate while maintaining interconnected voids, an increase in water/cement ratios has been reported to result in enhanced strength and density (Joshi and Dave, 2017). However, this also leads to a reduction in permeability. A water/cement ratio in the range

Figure 3.10 Influence of water/cement ratio on the strength of mortar at varying cement/sand ratio (Singh *et al.*, 2015b). Courtesy of Elsevier

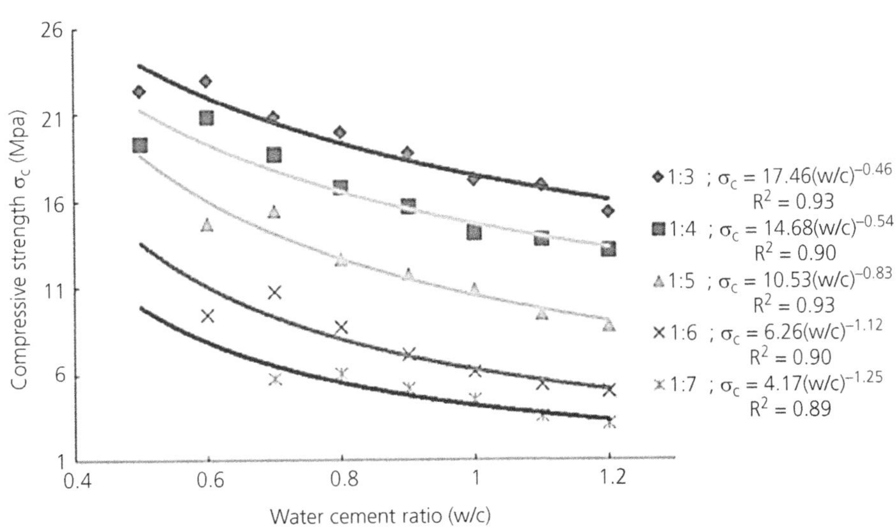

of 0.42–0.45 for complete hydration as well as the formation of sufficient hydrates to fill the void space and impart strength and durability in the case of HSC at a water/cement ratio of 0.35 has been suggested (CCAA, 2020). The development of superplasticisers now enables the production of UHSC with a very low water/binder ratio of less than 0.2, achieving adequate workability and slump. However, this approach recommends the use of low-reactive cement (Aïtcin, 2019). As discussed in the previous chapter, at the early stages of hydration with a higher water/binder ratio, the microstructure develops with more ettringite and portlandite than C-S-H due to the dissolution of C_3A. This negatively affects the bond and porosity at ITZ between HCP and aggregate, thus reducing the durability of concrete. The substantial influence of the water/cement ratio on the porosity and bond strength of ITZ has also been reported (Bentz *et al.*, 1992; Prokopski and Langier, 2000). The impact of varying water/binder ratios on concrete strength at different SF replacement levels, with superplasticiser content held constant, was reported (Bhanja and Sengupta, 2003) with a modified (relative to Abrams' formula) strength and water/binder ratio relationship, as shown in Figure 3.10.

Workability

The workability of concrete is expressed by fluidity, compatibility, and consistency that determines the ease of placement without undue segregation. In the case of workable concrete, it is possible to have the desired complete compaction by eliminating entrapped air to achieve a dense, homogeneous concrete with reduced porosity, and enhanced durability and strength. In situations involving very narrow spaces (such as narrow or micro trenches), in situ pile construction, or pumping concrete through pipes against gravity for high-rise buildings, highly workable concrete is essential. A compacted, dense concrete with lower porosity is unlikely to be affected by any external physical or chemical degrading agent.

Depending on the size of aggregate used, the workability of fresh concrete in British and European standards (different series of BS EN 12350, BSI (2019b)) is determined by slump test, vebe test, and flow table test. In the case of the slump test, the fresh concrete is compacted into a mould in the shape of a frustum of a cone following the specified procedure. When the cone is withdrawn upwards, the distance (in mm) the concrete has slumped provides the consistency or slump (S) of the concrete. In British standards, the slump values are classified as S1 to S5, where S1 is a relatively dry mix with slump values 10–40 mm and S5 is the most workable mix with 200–210 mm slump. The slump test is also performed routinely at laboratory and construction sites for quality control of the concrete.

Concrete workability is largely dependent on the water/binder ratio and, for a given water/binder ratio, concrete workability depends on the aggregate type and CA/FA ratio; an optimised ratio needs to be determined for a given material. The surface texture (such as crushed angular shape) and higher water absorption (porosity) of aggregate reduces the workability. An aggregate mixture having $> 20\%$ coarse aggregate retained on a given sieve size with an aggregate spacing $< 600\,\mu m$ for $\approx 75\%$ of the aggregates and cross-sections where no coarse aggregate was present showed poor workability (Sokhansefat *et al.*, 2019). Accumulation of bleed water underneath the aggregate particles, which creates voids in the transition zone, was indicated by Holland and Detwiler (2000), thus lowering the mortar or concrete strength.

The BRE concrete mix design guide (BRE, 1997) requires lower water content in the case of uncrushed aggregates (gravel, shingles, etc.) than crushed aggregates for equal workability, assuming the smaller the maximum size of the aggregate, the higher the water content needed.

Segregation and bleeding of concrete

Segregation is the separation of coarse aggregates from mortar during or after transportation and placement of the concrete. An improperly designed non-cohesive concrete mixture containing excessive fines with a high slump or lack of fines with high water content is susceptible to segregation. During segregation, when mixed water separates from the binder and is driven up to the surface of the plastic concrete after placing, bleeding occurs. A segregated and bled concrete results in inhomogeneity, a local high-porosity surface, and honeycombing in hardened concrete. It is weak in such a way that fingernails can mark it; it has reduced durability and low strength and is susceptible to being degraded by degrading agents. A properly designed cohesive concrete may be segregated due to over-vibration. The susceptibility to segregation and bleeding due to improper vibration of concrete pavement containing spherical fly ash as supplementary cementitious materials (SCM) and associated durability issues during service life was reported by Taylor and Sadati (2021). Water in concrete, being the lightest constituent, tends to rise during the sedimentation process, leading to bleeding. This phenomenon is closely associated with plastic shrinkage and cracking as heavier particles, such as aggregates and binders, settle. It is crucial to maintain a controlled amount of water on the exposed surface of the concrete in its plastic state to ensure that the evaporation rate does not exceed the rate of water migration to the surface. This balance helps mitigate drying and plastic shrinkage. Properly proportioned concrete mixes, incorporating water-reducing and air-entraining admixtures, can effectively minimise segregation and bleeding, thereby enhancing overall performance.

Admixtures for concrete

Unlike additions, admixture is a chemical added during the concrete, mortar, or grout mixing in a quantity not more than 5% by mass of the binder content which alters the properties of the mix in the fresh and/or hardened state for physical and economic advantages. Material engineers have been using admixtures in concrete for various applications from ancient times to modern-day construction. In addition to enhancing the durability of concrete, admixtures play a crucial role in reducing cement or binder content, thereby lowering the carbon footprint. The challenges posed by climate change, combined with the demand for high durability, safety, and aesthetic standards in modern infrastructure, have led material engineers to increasingly rely on advanced admixtures, especially in last 30 years. Modern sophisticated admixtures enable the production of high-performing concrete that meets stringent durability and performance requirements essential for modern infrastructure. It is estimated that in the UK more than 85% of all precast, ready-mixed and in situ concrete contains admixture (Dransfield, 2009).

Admixtures used in concrete, mortar, or grout can originate from chemical, mineral, or natural sources. The specifications and conformity criteria for these admixtures are outlined in the BS EN 934 series, consisting of six parts, with Part 2 (BSI, 2012b) specifically addressing admixtures for concrete. BS EN 934 lists 12 types of concrete admixtures, as shown in Table 3.1, along with their definitions and performance requirements. The table clearly indicates that admixtures are classed based on their function in the concrete mix, with some serving multiple roles, rather than being categorised by their chemical composition. In the following sections, some of the most commonly used admixtures are discussed, including their specific roles, benefits, and performance in concrete applications.

Water-reducing or plasticising admixtures

As discussed earlier, the compressive strength of concrete is inversely proportional to the water/cement ratio. For increased workability as well as to reduce the water/binder ratio for a given flow

Table 3.1 Types of admixtures listed in BS EN 934-2 (BSI, 2012b)

Name of admixture
Water reducing/plasticising admixture
High range water reducing/superplasticising admixture
Water retaining admixture
Air entraining admixture
Set accelerating admixture
Hardening accelerating admixture
Set retarding admixture
Water resisting admixture
Set retarding/water reducing/plasticising admixture
Set retarding/high range water reducing/superplasticising admixture
Set accelerating/water reducing/plasticising admixture
Viscosity modifying admixture

and for an improved paste-to-aggregate interfacial bond, high water-reducing admixtures (HWRA) are widely used in the construction industry (Xu and Beaudoin, 2000). However, while a very low water/cement ratio can help achieve higher strength, it significantly reduces the workability of the concrete. This reduced workability can lead to poor compaction, resulting in honeycombing and large pores. In applications like high-rise buildings or in situ deep pile construction, high workability, high-strength concrete, and low water/cement ratio are critical. In such cases, the use of superplasticising dispersing admixtures is crucial to ensure adequate workability for easy placement without compromising strength or durability. When substantial or high-range workability is required, then superplasticiser or high-range water-reducing admixture (HRWRA) is used instead of a plasticiser. These are sulfonated melamine formaldehyde condensates, sulfonated naphthalene formaldehyde condensates, and polycarboxylate ethers. Partial replacement of cement with silica fume synthesised with polycarboxylate ether-based HWRA was found to provide remarkable improvement in previous studies (Heikal *et al.*, 2006, Plank *et al.*, 2009, Toutanji and Elkorchi, 1995). Deflocculation and dispersion of hydrated cement particles by hydrophilic surfactants of HWRA or superplasticiser (SP) have been reported by Edmeades and Hewlett (2003). Generally, plasticiser allows around 10% water reduction, whereas in the case of superplasticiser 15–30% reduction is achievable. Plasticising admixture allows a reduction in the water content of a given concrete mix without affecting its workability. Likewise, this type of admixture increases the workability/slump of a given concrete without increasing the water content or changing any mix parameter.

Plasticisers, when added to concrete, mortar, or grout, adsorb onto the surface of cement or binder particles, imparting similar electric charges that cause the particles to repel each other. This repulsion leads to enhanced dispersion, which in turn increases the fluidity and workability of the mix. Plasticisers are long-chain polymers, either lignosulfonates, which are derived in the processing of wood for paper pulp, or polycarboxylate ether, whereas superplasticisers are based on sulfonated melamine formaldehyde condensates or sulfonated naphthalene formaldehyde condensates. The application of plasticisers offers numerous benefits, including enhanced strength,

Figure 3.11 Effect of superplasticisers on the fluidity of grout (Zhu *et al.*, 2023)

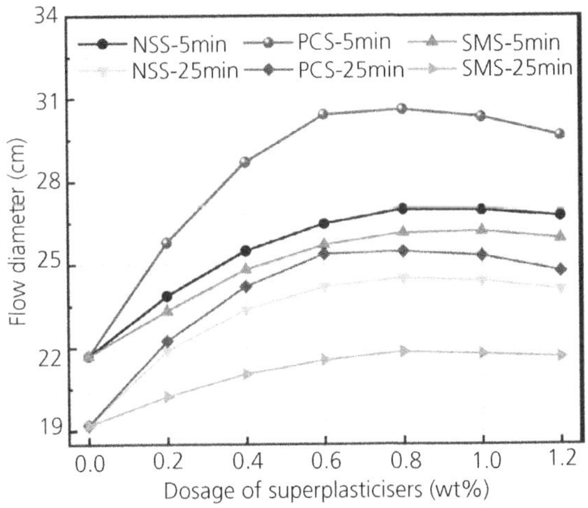

reduced permeability, minimised bleeding, self-compacting properties, improved surface finish for exposed concrete, ease of placement in densely reinforced areas, and increased pumpability. Figure 3.11 shows the effect of different dosages of naphthalene sulfonate (NSS), polycarboxylate (PCS), and sulfonated melamine superplasticiser (SMS) on grouts of the same water/binder ratio of 0.65 over time (Zhu *et al.*, 2023).

Set-retarding admixtures

This type of admixture prolongs the setting time of cementitious binders by slowing down the hydration reaction in the initial hours. While it has little impact on overall strength, retarding the set time is crucial for large concrete pours to prevent cold joints or discontinuities between batches. It is especially important when concreting in hot weather, where extended workability is needed for long hauls, or in slipform construction. Retarders work by adsorbing onto the binder particles, reducing their solubility and slowing the hydration process.

Common retarders include salts of lignosulfonic acids, hydrocarboxylic acids, and carbohydrates, as well as lead and zinc oxides, phosphates, magnesium salts, fluorates, and borates. Sucrose and citric acid are also used as effective retarders. The effect of increasing borax content from 2% to 8% on the flow and setting time of alkali-activated binders, at a water/binder ratio of 0.3, was reported by Oderji *et al.* (2019), as illustrated in Figure 3.12.

Set-accelerating admixtures

Accelerators are used to enhance the setting or hardening of concrete, mortar, and grout. BS EN 934-2 (BSI, 2012b) classifies these admixtures into set accelerators and hardening accelerators. Set accelerators shorten the time it takes for the mixture to transition from a plastic to a rigid state, while hardening accelerators enhance strength development, particularly within the first one or two days, with or without affecting the setting time. Typically, calcium nitrate and sodium thiocyanate are used as set and hardening accelerators, respectively. Calcium chloride was once commonly used for both purposes; however, due to the corrosive effects of chloride on reinforcement, the use of chloride-based accelerators has been largely discontinued.

Figure 3.12 Influence of different dosages (2% to 8%) of borax as retarder on alkali-activated binder (Oderji *et al.*, 2019)

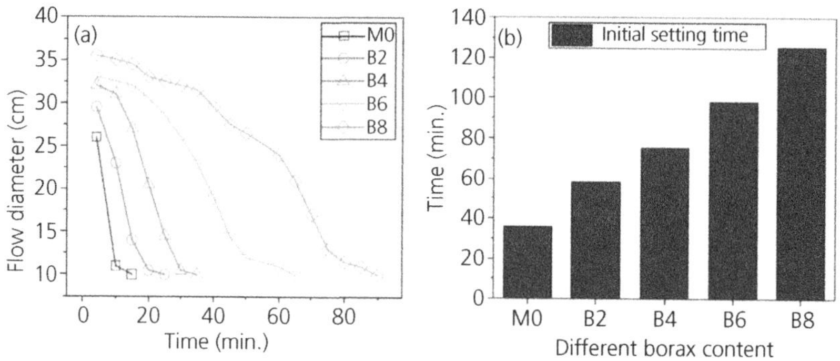

Accelerators are essential during emergency repair when early formwork removal is necessary or when concreting in cold weather, where prolonged curing times are required. Set accelerators reduce both the initial and final setting times, allowing for quicker finishing in cold conditions. Additionally, they can reduce total bleed in the concrete. While hardening accelerators boost early strength by speeding up the hydration reaction, they may also lead to higher heat of hydration, increased drying shrinkage, and creep, as well as reduced long-term strength (CCAA, 2020). The rate of acceleration is influenced by the type and concentration of the accelerator, type of binder, binder content, and temperature.

Air-entraining admixtures (AEA)

When added during mixing, AEA introduce a controlled amount of small, uniformly distributed air bubbles that remain in the concrete after compaction and hardening. These entrained air bubbles differ from entrapped air, which occurs in inadequately compacted concrete. The intentionally created bubbles are spherical, range in size from 10 microns to 1 mm, and are not interconnected. As a result, the porosity of the hardened concrete does not increase noticeably, although there may be a slight reduction in strength. This impact can be offset through proper mix design. The primary purpose of AEA is to improve the freeze–thaw resistance of concrete, although their application in improving the plastic performance of concrete with low binder content and to compensate the workability resulting from poor gradation of sand are proven (CCAA, 2020). The influence of AEA on concrete strength and durability is illustrated in Figure 3.13 (CCAA, 2020).

Similar to plasticisers and superplasticisers, AEA adsorb onto cement particles, although they function through distinct mechanisms with different effects in the mix. The entrained air bubbles act as pressure relief points, mitigating the stresses caused by the expansion of trapped frozen water within the concrete. Common AEAs include Vinsol resin derived from pine wood, soaps of oleic and caproic acids, and synthetic alkyl sulfates. Several factors influence air content in concrete, including the type and dosage of AEA, sand gradation, binder type and content, mixing equipment, and temperature. It is important to note that the compatibility test for air-entraining agents in combination with other admixtures in a mix needs to be performed in advance. BS EN 206 (BSI, 2021c) specifies a minimum air content of 4% for concrete subjected to moderate to high exposure class.

Figure 3.13 Effect of air entrainment admixture on compressive strength and durability. Reproduced with permission from Cement Concrete & Aggregates Australia (CCAA, 2020, Part II, s. 5, p. 5, figure 5.3)

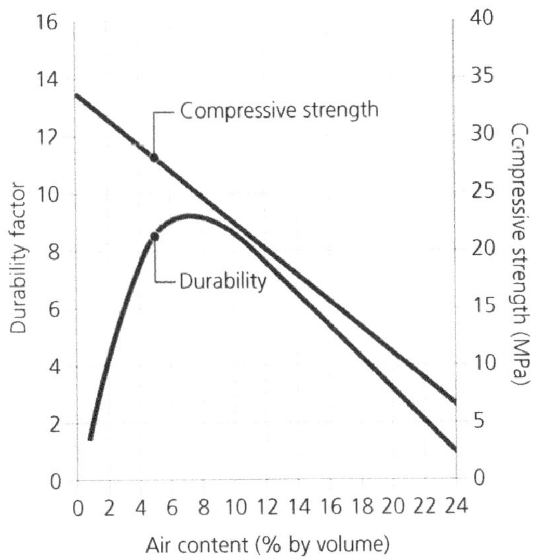

Admixtures and sustainability

The proper use of admixtures provides significant durability, economic, and sustainability benefits. For instance, accelerators promote rapid strength development in mixes with a high proportion of supplementary cementitious materials (SCMs), helping to address the challenge of low early strength due to the reduced amount of Portland clinker. Retarders, by minimising the risk of cold joints and discontinuities, contribute to the creation of dense, durable concrete structures with minimal permeability. Admixtures do not contain hazardous chemicals that require special health and safety protocols. Additionally, the risk of leaching of chemicals used in admixtures bound within hardened concrete is effectively zero, ensuring a safe and environmentally friendly performance over the concrete's lifespan.

While the embodied carbon associated with admixtures is relatively high, they account for less than 1% of the total embodied carbon of concrete due to their low dosage. At the same time, they enable reductions in other high-carbon components, such as cement content, section size, and materials. Admixtures enhance the durability of structures in various environmental conditions, resulting in reduced maintenance, extended service life, and lower overall lifecycle costs. The versatility of modern admixtures for their excellent contribution to flowability, retention, strength, microstructure modification, sustainability, and nano modifications was reported by Ghosal (2023). Allowing and encouraging the use of admixtures with demonstrated performance in concrete mixes with low Portland cement content is suggested in aspects of flexibility in concrete specification in the *Low Carbon Concrete Routemap* (Low Carbon Concrete Group, 2022).

Precast concrete

As the name suggests, precast concrete refers to prefabricated structural, architectural, or hollow-core concrete elements cast at a location other than the final installation site. Figure 3.14 shows that

Figure 3.14 Precast beam supported on precast concrete columns. Reproduced with permission from Cement Concrete & Aggregates Australia (CCAA, 2020, Part VI, s. 20, p. 10, figure 20.10)

these elements are commonly used in manufacturing structural and non-structural elements such as floor slabs, foundation piles, stairs, paver blocks, fencing, box culverts, beams, and bridge components. Precast concrete is typically manufactured in dedicated plants or factory settings and later transported to the project site for installation. The Mineral Products Association (MPA) Precast and MPA Masonry represent UK precast concrete manufacturers and, in 2022, out of 16.62 million tonnes, 14% (2.3 million tonnes) was used by architectural and structural precast concrete (MPA Precast and MPA Masonry, 2024).

The controlled production environment, ensuring precise mixing, compaction, and curing, enables superior quality compared to in situ concrete. This approach allows for the efficient use of materials with confidence, saving 23% of materials, particularly in thinner sections, which is a notable advantage (Shanks *et al.*, 2019). Additionally, reduced project duration and construction independent of weather conditions are notable benefits of precast concrete, essential for availing offsite construction opportunities or design for manufacture and assembly (DfMA) where reduced waste and energy are likely. However, currently, limited benefits from precast concrete due to the use of higher cement content without any SCM for rapid demoulding (for higher production) were noted and suggested a greater uptake of lower-carbon concretes within precast elements (Low Carbon Concrete Group, 2022). To achieve this, adjustments to the factory casting sequence were recommended, allowing precast elements to remain in the moulds for an extended period before removal. The use of SCMs in precast concrete products increased to 25.7% in 2022, up from 17.9% in 2020. However, a decline of approximately 8% in the use of recycled and secondary aggregates compared to 2012 was reported, primarily attributed to challenges in maintaining an adequate supply (MPA Precast and MPA Masonry, 2024). While precast concrete offers numerous advantages over cast in situ concrete, significant challenges include safety risks, logistical complexities, site storage requirements, and emissions associated with handling and transporting large precast elements.

The general common specifications for a large variety of precast concrete products manufactured in factory settings are covered by BS EN 13369 (BSI, 2023), with additional standards applicable for each type of structural element. This standard emphasises the appropriate use of concrete constituents, placing, and compaction avoiding segregation, with proper embedding of the reinforcement and curing to control drying out and shrinkage cracks. The constituents such as binder, aggregate, admixture, and additions used in precast concrete are controlled by BS EN 206 (BSI, 2021c).

Roller-compacted concrete (RCC)

As the name implies, RCC is an engineered zero-slump concrete compacted using roller-type mobile compacting machines, and which does not contain reinforcement or formwork. RCC contains similar constituents to conventional concrete but is typically used in applications such as concrete dams and pavement construction (Figure 3.15). RCC remains particularly suitable for projects requiring mass concrete of lower grades and is stiff enough to compact efficiently using mobile equipment. For applications involving the wearing course of pavement, the required skid resistance of RCC can be achieved through mechanical grooving and grinding or by treating the surface with an asphalt overlay. Due to its cost-effectiveness and ease of construction, the UK's National Highways updated its design standards in 2021 by replacing HD 26/06 with CD 226 (National Highways, 2021), with a provision for using RCC. This revision introduced RCC as a design option for rigid pavement base and binder courses. In such cases, RCC made with C40/50 concrete must achieve a flexural strength of 5 MPa, an elastic modulus of 50 000 MPa, and a Poisson's ratio of 0.2, where its in situ strength is largely influenced by the degree of compaction achieved after placing the concrete. The British Cementitious Paving Association (Britpave) has published a report for RCC construction methodology and techniques (Britpave, 2022), providing guidelines on materials selection, mix design, and quality control. A dense, well-graded aggregate containing 85% of the mixture can contain binders specified in BS EN 197-1 BSI (2019a) of the order of 270–300 kg/m^3.

The curing of RCC is critical to ensuring its overall durability, given its very low workability and zero-slump characteristics. The combination of low water content, high production rates, and

Figure 3.15 RCC application in pavement construction. Reproduced with permission from Britpave (Britpave, 2022)

large, flat, open working areas further underscores the importance of adequate curing measures to achieve optimal performance and long-term durability (Nanni, 1988). The influence of four types of curing methods using burlap, membrane, water, and curing compound for a duration ranging from 0–60 min after casting the RCC containing 350 kg/m^3 cement at 0.41 water/cement ratio was investigated by Keleş and Akpinar (2022). Water curing was found to be the best method and the best result was obtained from 60 min curing after casting.

Questions to practice

1. Explain the roles of C_3S and C_2S in the hydration reaction of cement.
2. Define ITZ and its influence on concrete strength.
3. Briefly explain how the water/binder ratio influences concrete strength at the micro-structural level.
4. Justify the use of an appropriate admixture considering the influence of water/binder ratio on concrete strength and workability.
5. Recommend a suitable admixture for concrete used in constructing a thick foundation with a large pour, providing justification for its selection.

Monower Sadique
ISBN 978-1-83608-477-8
https://doi.org/10.1108/978-1-83608-474-720251004

Chapter 4
Concrete degradation

After completing this chapter, you should be able to

- understand the concepts of quality and durability in concrete and their significance
- identify the mechanisms through which concrete degrades due to physical and chemical actions
- explore the sulphate attack mechanism in concrete, its effects, and strategies for control and prevention
- analyse the impact of chloride ingress on concrete and evaluate measures to mitigate this degradation
- understand the carbonation process in concrete, its effects on durability, and measures to control it
- identify the sources and effects of acid attack on concrete, along with strategies for protection and durability enhancement
- explore the alkali–silica reaction (ASR) mechanism in concrete, its effects, and methods to minimise its impact
- examine the degradation of concrete due to fire exposure, its impact on structural integrity, and control measures
- investigate the effects of freeze–thaw cycles on concrete, and the mix design parameters necessary to enhance resistance.

Introduction

The performance of in-service concrete under environmental exposure depends primarily on its quality, durability, and assigned exposure class. Concrete degradation can be mitigated by enhancing durability (e.g. increasing impermeability) or managing exposure to harmful agents (e.g. reducing CO_2 levels or remediating contaminated soil). Key properties affecting degradation include concrete porosity and the relative humidity of the exposure environment, both of which impact the bond between the cement paste and aggregate, potentially leading to cracking and spalling. Excess water added to improve concrete workability is not chemically utilised in the hydration reaction, leading to pore formation in the hardened concrete. These pores increase susceptibility to degradation from external harmful agents. Although concrete is a resilient material, it is degraded by sustained action of various environmental conditions such as chlorides, sulphates, fire, and so on. Based on the degrading mechanism, various controlling measures in terms of selective constituent materials, limiting values of mix design parameters, and so on are specified in standards.

Quality and durability of concrete

The quality and durability of concrete are inseparable parameters. Although concrete quality is heavily dependent on its constituents, good quality concrete should exceed the required strength, have a binder content that meets or exceeds specifications, maintain a water/binder ratio within specified limits with the desired consistency in its fresh state, and be well cured. Durable concrete should be able to remain in service safely for at least the designed lifespan under existing structural and environmental loads.

The required quality of a particular concrete to ensure durability depends on several factors, with the most important being the design life of the structure, structural requirements specified by the respective design codes, expected environmental conditions (such as the risk of corrosion, freeze–thaw cycles, or sulphate attack, etc.), exposure to any aggressive environments (such as sulphate-contaminated ground or marine environment with strong wave action, etc.), and minimum cover to reinforcement.

Designing durable concrete begins at the concept design stage and continues through the design, detailed specification, and execution phases. Achieving durability in practice also requires that maintenance is carried out as planned during the design stage. In the UK, BS 8500 (BSI, 2021a, 2021b) provides guidance and limiting values for the above parameters on the concrete quality and durability provisions. Although concrete is considered a highly durable material compared to other construction materials, degradation of concrete during its service life may occur through chemical (chloride, sulphate, acid, etc.) and physical (fire, freeze–thaw, etc.) actions on concrete due to environmental factors to which the concrete is exposed or internal factors such as contaminated aggregates. In all cases, the rate of degradation is principally influenced by the permeability and/or porosity of concrete as well as the concentration and/or scale of harmful degrading agents.

Concrete degradation through chemical actions

Generally, well-designed, cured, and constructed concrete performs satisfactorily during its service life under most environmental conditions. However, its intended service life can be significantly reduced when exposed to certain chemical environments if specific protective measures are not incorporated during the design and construction phases. It is worth mentioning that concrete is rarely attacked by dry chemical agents and a significant concentration of chemicals in solution need to come into close contact with concrete either internally or externally for any degradation to occur. Some common chemical agents that initially attack concrete and degrade its integrity are sulphates, chlorides, carbon dioxide, acid, and alkali–silica. The following sections briefly describe the source of these agents, their degradation mechanisms, and controlling measures, including provisions in standards.

Sulphate degradation and controlling measures

Sulphate, a naturally occurring mineral, is one of the most common degrading agents that attacks cement-based materials. Although a controlled amount of sulphate-bearing material (less than 5%), such as gypsum, is added as a grinding aid and for setting control during clinker production, excess sulphate can be harmful. Common sources of sulphate exposure for concrete include contaminated soil, groundwater, industrial effluents, and fertilisers. In the UK, aggregates with sulphate contamination are rare. However, globally, some natural aggregates are quarried from sources with sulphate levels high enough to insidiously degrade concrete. During 1945–1970 in the UK, thousands of domestic properties were built using sulphate-bearing fill materials originating from colliery spoil and industrial by-products to support concrete ground floor slabs without using any separating dampproof membrane. Degradation was found to be most severe in front of the fireplace, where sulphate-bearing solutions from the underlying hardcore were drawn up through the slab by the heat generated from the fire (BRE, 2008). Figure 4.1 shows a typical concrete wall damaged through cracking as sulphate-bearing contaminated water attacked the concrete.

Building foundations, the underside of ground-bearing concrete slabs, and retaining walls are most vulnerable to sulphate attack in the case of contaminated settings. Although the degradation mechanism through sulphate attack is very complex, it is widely accepted that the solution of calcium, sodium, potassium, or magnesium sulphates when present in sufficient concentration attack the hydrated C_3A phase in HCP and form expansive alkali sulfo-aluminate hydrate or ettringite in the form of monosulphate (AF_m). Softening and dissolution of HCP in hardened concrete is also suggested when sulphates attack and remove calcium hydroxide (Concrete Society Working Party, 2000). In the case of attack by calcium sulphate, calcium sulfo-aluminate $3CaO \cdot Al_2O_3 \cdot 3CaSO_4 \cdot 31H_2O$ forms which causes expansion, causing cracking and spalling followed by complete disintegration or loss of bond between HCP and aggregate.

The rate of degradation due to sulphate attack is influenced by several factors, including the concentration of sulphates, the C_3A content of the clinker, and the quality and permeability of the concrete. According to BS EN 206 (BSI, 2021c, table 2), exposure classes are categorised based on the concentration of sulphates present in natural soil and groundwater into slight, moderate, and highly aggressive environments, designated as XA1, XA2, and XA3, respectively, providing limiting values for concrete mix design parameters, such as maximum water/cement ratio, minimum strength class, and minimum cement content. Since higher C_3A content in clinker increases the vulnerability to sulphate attack, using clinker with low C_3A content, such as sulphate-resisting Portland cement (SRPC), is recommended as a primary measure. Additionally, applying impervious coatings to concrete exposed to highly aggressive environments can further protect against sulphate ingress and degradation.

Figure 4.1 Sulphate attack of a wall in contact with sulphate-contaminated water (Mindess, 2019). Courtesy of Elsevier

Chloride degradation and controlling measures

Common sources of chloride exposure for concrete include de-icing salts spread on bridge decks and car parks in winter (as shown in Figure 4.2), marine structures in coastal areas, water treatment plants, swimming pools, chloride-containing admixtures (such as accelerators), and contaminated aggregates or water used in the concrete mix. The most vulnerable component to chloride attack is the reinforcement embedded in concrete, which can lead to corrosion and compromise structural integrity.

Surrounding the steel with an alkaline environment having a pH value within the range of 9.5 to 13 to control corrosion is a proven technique. The steel reinforcement inside concrete is electrochemically protected against corrosion thanks to the alkaline environment (passive layer) provided by cement in concrete. Hydrated cement containing calcium hydroxide as well as a controlled amount of sodium and potassium in the pore solution provides such an environment. This means that steel embedded in well-cured, high-quality concrete is electrochemically protected from corrosion, provided there is adequate concrete cover and no unforeseen aggressive chemicals are present that were not accounted for during the design phase. At this higher pH value (> 12.5), a passive film forms on the steel that reduces the rate of corrosion to a very low and harmless value. Thus, the concrete cover provides chemical as well as physical protection to the steel. In uncontaminated concrete, this passive layer is continuously broken and immediately replenished. Depassivation and consequently the corrosion of the reinforcement is initiated either when the pH within the concrete drops and/or the rate of replenishment fails to match the rate of breakdown of the passive layer. Chloride has the capacity to destabilise the breaking down and reinstatement of the passive layer without lowering the pH. The concentration of chloride in the pore solution or the threshold value at which depassivation is triggered is not straightforward. Generally, this threshold value of chloride content was reported to vary between 0.02% and 3.08% per weight of the binder. However, the ratio of Cl^-/OH^- and pH also influences the rate of depassivation.

BS EN 206 (BSI, 2021c) classifies the exposure classes for risk of reinforcement corrosion caused by chlorides other than from seawater as XD1 (moderate humidity), XD2 (wet, rarely dry), and XD3 (cyclic wet and dry). The most severe condition that leads to rapid chloride penetration is cyclic wet and dry (XD3), when 80–98% ambient relative humidity exists, such as parts of bridges

Figure 4.2 Chloride-induced corrosion caused by de-icing salts in a concrete car park (author's own)

exposed to spray containing chlorides, pavements and car park slabs, or the surface of marine structures exposed to high and low tides (bridge pier, sea wall, etc.).

Restricting the seepage of water and oxygen into concrete is the most fundamental approach to control the chloride attack on concrete. In achieving this, a well-cured and well-compacted good quality concrete containing additions (SCM) with low permeability and sufficient clear cover are essential. Accordingly, BS EN 206 (BSI, 2021c, table X) specifies the maximum water/cement ratio and minimum cement content and strength class based on the exposure class to control the chloride attack. Cement with low C_3A content was found to control chloride degradation by binding a portion of chloride by forming calcium chloroaluminates (CCAA, 2020).

Carbonation degradation and controlling measures

As discussed in the previous section, corrosion of reinforcement in concrete as a consequence of depassivation occurs due to a reduction of pH. Dissolution of CO_2 from the atmosphere in concrete pore solution reacts with C-H and C-S-H present in HCP, initially forming carbonic acid which further forms calcite ($CaCO_3$) and neutralises the pH to 9, as shown in Equations 4.1 and 4.2. The depassivation of steel through the neutralisation of pH that initiates corrosion through CO_2 attack is called carbonation. Figure 4.3 shows carbonation degradation observed on a 25-year-old retaining wall adjacent to a busy road.

Figure 4.3 Carbonation-induced corrosion in a concrete retaining wall adjacent to a busy road (author's own)

$$H_2O + CO_2 \rightarrow H_2CO_3 \tag{4.1}$$

$$H_2CO_3 + Ca(OH)_2 \rightarrow CaCO_3 + 2H_2O \tag{4.2}$$

The above reaction only occurs in solution; hence in dry concrete, the risk of carbonation degradation is low although the rate of CO_2 ingress in dry concrete will be faster than in wet concrete due to reduced diffusion through saturated pores. Carbonation starts at the surface and penetrates slowly into the concrete. In carbonated concrete, corrosion of reinforcement is triggered once the carbonation front (where pH drops to 9 or below) reaches the reinforcement.

Equation 4.3 shows the most well-established relationship between carbonation depth (x in mm) and time (t in year) or duration of exposure which is widely used for assessing the risk of corrosion during structural health monitoring of concrete

$$x = k\sqrt{t} \tag{4.3}$$

where k is the carbonation constant related to the environment, the concentration of CO_2, exposure class, and concrete quality.

Carbonation-related degradation in concrete is primarily influenced by environmental conditions and concrete quality. For instance, with a water/cement ratio of 0.60, carbonation can reach a depth of 15 mm in 15 years, whereas with a ratio of 0.45, the same depth may take 100 years (Neville, 1996). Additionally, inadequate curing (dry curing) and higher water/cement ratios have been shown to significantly reduce concrete's resistance to carbonation (Ben-Bassat *et al.*, 1990).

The carbonation constant of a particular concrete structure with known exposure can easily be measured by using a phenolphthalein indicator on the extracted concrete sample. The phenolphthalein indicator appears pink in contact with alkaline concrete with pH values more than 9 and colourless at lower levels of pH, as shown in Figure 4.4. Due to the absence of free lime in high-alumina cement, a phenolphthalein test cannot be performed.

BS EN 206 (BSI, 2021c) classifies the exposure classes for risk of reinforcement corrosion caused by carbonation as XC1 (dry or permanently wet), XC2 (wet, rarely dry), XC3 (moderate humidity), and XC4 (cyclic wet and dry). The most severe condition that leads to rapid carbonation is cyclic wet and dry (XC4) when 60–80% ambient relative humidity exists. Similar to controlling chloride, restricting the diffusion of CO_2 into the concrete is the most fundamental approach to controlling the carbonation of concrete. In achieving this, a well-cured and well-compacted good quality concrete containing additions (SCM) with low permeability and sufficient clear cover is essential. Accordingly, BS EN 206 (BSI, 2021c, table X) specifies the maximum water/cement ratio and minimum cement content and strength class based on the exposure class to control carbonation-induced degradation.

Acid degradation and controlling measures

Alkaline HCP is readily vulnerable to attack by any acid in liquid form, although concrete containers can be used to store acid in dry form. Common sources of acidic solution exposure for concrete include acid rain which contains dilute solution carbon dioxide (CO_2) and sulphur dioxide (SO_2) in industrial regions, CO_2- and SO_2-bearing water resulting from organic decomposition in marshy

Figure 4.4 Application of phenolphthalein indicator to measure carbonation depth (author's own)

and peaty areas, sulfuric acid generated due to bacterial decomposition of sewage in sewer lines and near the stockpiles of mine tailings. Like carbonation, acid attacks the calcium hydroxide (C-H) and calcium silicate hydrate (C-S-H) of HCP and converts them to calcium salts of the acid, which are relatively soluble and leach out from the concrete and destabilise the bonding between HCP and aggregate, as shown in Figure 4.5.

High-quality, well-cured concrete with additions where accelerated pozzolanic reactions can lower the C-H content in the HCP can provide resistance to acid attack. In the case of highly aggressive

Figure 4.5 Sulfuric acid degradation in concrete sewer line (PCA, 2002)

acidic conditions (such as industrial floors), coating the concrete with epoxy resin and treating the surface with diluted water glass (sodium silicate) solution is widely suggested.

Alkali–silica reaction and controlling measures

Cement clinker contains a controlled amount of alkali oxides (sodium oxide and potassium oxide). A reaction can occur between these alkalis and reactive silica originating from contaminated aggregates such as chert, opal, and so on, leading to the formation of silica gel which is expansive and hygroscopic at the ITZ. This reaction is called an ASR or alkali–aggregate reaction. The hygroscopic reaction product (gel) swells in the presence of moisture, leading to widespread map-like cracking across the concrete surface, as shown in Figure 4.6. Three conditions, namely contaminated aggregate containing reactive silica; concrete pore solution with high alkali content; and sufficient amount of moisture need to be met for ASR to occur, as shown in the following schematic equation.

$$(Na/K)OH + SiO_2 + H_2O \rightarrow Na_2SiO_3 \cdot H_2O \quad (4.4)$$

Alkali–silica water alkali–silica gel (expansion)

Since the root cause of this degradation is the contaminated aggregate, the damage occurs insidiously, triggered by the reaction between concrete constituents (cement and aggregate) when exposed to external moisture. The most widely adopted controlling measures against ASR are using non-reactive aggregate tested by petrographic analysis or sourcing aggregate from a quarry that has no record of silica contamination, using cement with low alkali content (Na_2O equivalent lower than 0.6%) and using low water/cement ratio in addition to using certain additions (silica fume, meta-kaolin) in the concrete mix.

Figure 4.6 Typical map-like cracking for ASR (PCA, 2002)

Concrete degradation through physical actions

As discussed, well-designed, properly cured, and well-constructed concrete performs reliably under substantial loads, and exposure to weather, fire, abrasion, and other physical stresses. Abrasion can cause surface wear on concrete structures like bridge decks, seawalls, and canal locks. In reinforced concrete, severe abrasion may reduce the clear cover, compromising durability and structural integrity. Concrete performance can be significantly compromised when subjected to prolonged exposure to certain physical conditions, particularly fire and frost, which are critical factors to consider. The degradation mechanism and controlling measures for concrete subjected to fire and frost are discussed in the following sections.

Fire degradation and controlling measures

Concrete is non-combustible and does not release toxic gases when exposed to fire. Its low thermal conductivity makes it a versatile and widely preferred construction material compared to alternatives. However, when exposed to prolonged high temperatures, the fire resistance of concrete structural and non-structural elements diminishes, eventually leading to compromised structural integrity and safety. Various parts of BS 476 specify fire tests for materials and fire resistance tests for building elements, with BS 476-10 (BSI, 2009a) outlining the principles, selection, role, and application of fire testing methods and their outputs. A building's responses against various parameters such as ease of ignition, the spread of fire, release of smoke, heat flow, collapse under self-weight, and melting represent a risk to life. Although it is hard to estimate, the rate and level of degradation of a concrete element depend on the rate of spread, the highest temperature reached, duration of exposure, concrete constituent, moisture content, and size of the element. Preserving the structural capacity in the event of fire of buildings with high fire hazards such as tunnel linings, parking, and warehouse storing of explosive chemicals is always of interest to materials and structural engineers. The severe fire at a car park in Liverpool, UK, on 31 December 2017 completely gutted the seven-storey concrete structure, destroying up to 1400 vehicles. Although the building had a substantial reinforced concrete frame that largely withstood and contained the fire, there were reports of some concrete floor collapses, as shown in Figure 4.7. Figure 4.7 illustrates that the reinforced beams and columns sustained only minor surface damage, while the precast ribbed floors disintegrated and collapsed under the intense heat. Following the incident, the Standing Committee on Structural Safety (SCOSS) published an alert report, classifying the event as a near-miss in terms of both life safety and structural integrity (SCOSS, 2018).

Concrete's inherent thermal properties start to deteriorate when exposed to prolonged fire due to several factors: moisture loss; degradation of the HCP and aggregates; the differing thermal behaviour of HCP and aggregates; and pore pressure generated by trapped moisture under sustained high temperatures. The higher risk of spalling for concrete over 60 MPa compressive strength containing silica fume and particular aggregate is reported by MPA The Concrete Centre (2019). Atkins (2015) reported discolouration in structurally fire-damaged concrete, attributed to ferrous salts present in the aggregate. The damage process of concrete under fire is complex; Soutsos *et al.* (2017) identified the following three key contributors to this degradation.

- Evaporation. The evaporation of moisture within concrete begins at around 100°C; as the temperature rises, tightly bound moisture is progressively driven off. In NSC, which has higher porosity compared to HSC, this moisture can escape more easily. However, in HSC, the denser structure traps the evaporated moisture, generating significant pore pressure. This buildup of pressure can lead to explosive spalling, where the concrete surface is violently delaminated, which can happen immediately after the fire or at later stages even after the fire

Figure 4.7 The scale of destruction and concrete disintegration after a fire at a multi-storey car park. Reproduced with permission from Merseyside Fire and Rescue Service

has been extinguished. Typical spalling of a fire-damaged concrete column is shown in Figure 4.8. Reduced mechanical resistance to fire has been reported for HSC when exposed to fire for durations exceeding 160 minutes, compared to NSC. To address this issue in tall buildings, the use of fireproof coatings or increasing the cross-sectional dimensions of HSC elements has been recommended (Dung and Unluer, 2017).

- Differential thermal expansion. At around 500°C a number of physical and chemical changes occur within concrete in addition to induced thermal stress due to disparity in thermal expansion between the HCP and the aggregate, particularly at the ITZ, leading to cracking and rapid strength loss. The mismatch in thermal expansion becomes even more pronounced above 600°C when coarse aggregates continue to expand while the C-S-H gel shrinks due to the loss of both physically and chemically bound moisture, as reported by Kore Sudarshan and Vyas (2019).
- Breakdown. At approximately 1000°C, the hydrates in the HCP completely disintegrate, causing concrete to lose its strength and structural capacity entirely. Up to 500°C, the strength loss is gradual, but beyond this point, the deterioration accelerates, culminating in total failure at or above 1000°C. Although the rate of increase of temperature through concrete elements is relatively slow, if the reinforcement is exposed to fire, either due to spalling of the concrete or prolonged heat exposure, the temperature of the reinforcement rises quickly, further accelerating the strength loss and compromising the structural integrity of the elements.

Figure 4.8 Spalling of a fire-damaged concrete column exposing reinforcement. Reproduced with permission from Sandberg LLP

The fire classification of construction materials and testing is outlined in BS EN 13501-1 (BSI, 2018), which categorises materials into seven grades, with A1 being the highest designation for incombustible materials. Concrete and masonry are classified as A1. Given that aggregates occupy a significant volume of concrete, selecting fire-resistant aggregates is a crucial consideration for materials engineers. Lightweight aggregates and limestone, which exhibit lower thermal expansion, are expected to offer superior fire performance compared to siliceous aggregates.

It is important to recognise that the fire performance of concrete refers to the performance of a specific structural element, rather than the concrete material alone. Therefore, when designing fire-resistant and efficient structures, two critical design considerations are emphasised: the overall dimensions of the structural elements and the average concrete cover over the reinforcement (MPA The Concrete Centre, 2019). These factors are critical in enhancing the fire endurance and structural integrity of concrete elements during fire exposure. Guidance on nominal cover to reinforcement for structural and fire consideration is also available in respective parts of BS EN 13369 (BSI, 2023).

Freeze–thaw degradation and controlling measures

In cold climates, concrete that is saturated or partially saturated and subjected to repeated freeze–thaw cycles experiences significant degradation. When water within the porous concrete freezes, it expands by approximately 9%, requiring additional space within the HCP. If no extra space is available, the expanding frozen moisture generates internal hydraulic pressure and creates microcracks, which are released on thawing. Successive repeated freeze–thaw cycles can exceed the strain tolerance of the concrete, leading to progressive damage, such as crumbling, cracking, and spalling.

Given this degradation mechanism, it is evident that dry concrete is less susceptible to frost damage. Although there is no specific threshold for saturation, being subjected to frequent splashing with water containing de-icing agents and exposed to freezing are classified as the most severe exposure (XF4) under BS 8500-1 (BSI, 2021a). Concrete structures such as bridge parapets, bridge piers near the roadside, road kerbs, and elements in hydraulic structures just above the water level are vulnerable to being affected by frost. A frost-damaged railway platform under frequent use of de-icing salt in winter can be seen in Figure 4.9.

Generally, frost-related damage is primarily associated with HCP, but the porosity of aggregates also plays a significant role. Aggregates with very high porosity allow hydraulic pressure to dissipate, while those with very low porosity do not retain enough water to generate pressure and are thus less susceptible to damage. However, aggregates with intermediate porosity, particularly those with total water absorption greater than 1.5%, are the most vulnerable to frost damage (Pigeon and Pleau, 2010; Verbeck and Landgren, 1960). These aggregates become saturated with seepage water but do not permit the release of pressure caused by ice crystal formation, leading to internal stress and potential damage followed by popout, as shown in Figure 4.10.

As discussed in the admixture section, AEA are effective in enhancing the freeze–thaw resistance of concrete by introducing entrained air bubbles within the HCP, which act as pressure release points (as shown in Figure 4.11). The superior performance of air-entrained concrete against frost damage has been reported widely (Soutsos et al., 2017; Tunstall *et al.*, 2021). BS 8500-1 (BSI, 2021a) specifies limiting values for minimum air and cement content, associated with maximum aggregate sizes, based on exposure class, to ensure the frost resistance of air-entrained concrete. In addition to using the appropriate admixture at an optimised dosage, well-cured concrete with low permeability and absorption, along with high-quality cement mortar, is critical for frost resistance. Aggregates with a history of poor performance in frost-prone environments should be avoided to further mitigate the risk of damage.

Figure 4.9 Frost-damaged concrete railway platform subjected to de-icing salt (author's own)

Figure 4.10 Aggregate produces a popout after a frost attack (PCA, 2002)

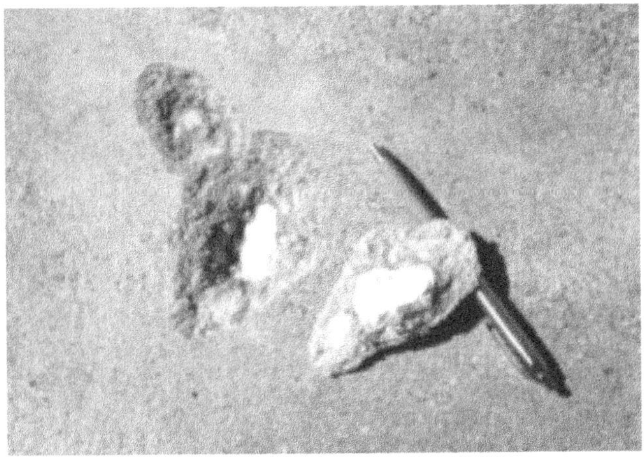

Figure 4.11 Air void shells in a cement mortar air entrained with AEA (Tunstall *et al.*, 2021). Reproduced with permission from Elsevier

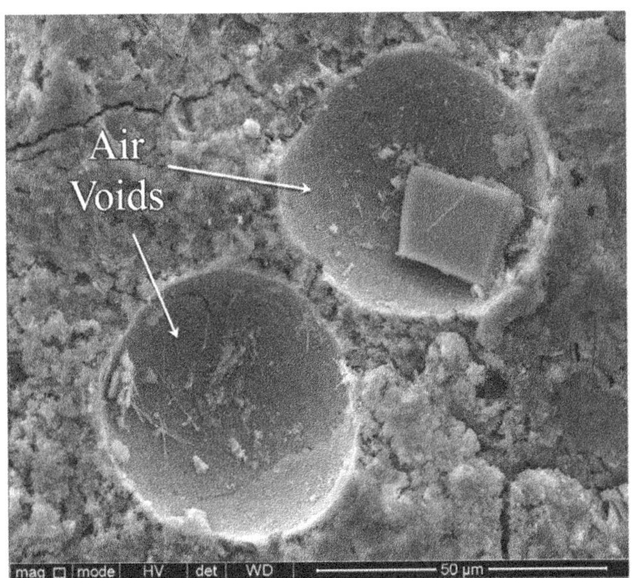

Concrete exposure class and mix design parameters

BS EN 206 (BSI, 2021c) classifies various exposure classes based on the environmental (chemical and physical) actions on concrete along with a description of the environment for each class. The environmental actions are corrosion by carbonation, corrosion induced by chloride from seawater and non-seawater, freeze–thaw attack, and chemical attack. Based on the severity of exposure, limiting values for concrete mix design parameters (binder content, water/binder ratio, strength class, etc.) have been suggested, as shown in Table 4.1.

Table 4.1 BS EN 206 recommended limiting values for concrete composition in relation to exposure class (BSI, 2021c)

Table F.1 — Recommended limiting values for composition and properties of concrete

	Exposure classes																	
	No risk of corrosion or attack	Carbonation-induced corrosion				Chloride-induced corrosion						Freeze/thaw attack				Aggressive chemical environments		
						Sea water			Chloride other than from sea water									
	X0	XC 1	XC 2	XC 3	XC 4	XS 1	XS 2	XS 3	XD 1	XD 2	XD 3	XF 1	XF 2	XF 3	XF 4	XA 1	XA 2	XA 3
Maximum *w/c*[c]	–	0,65	0,60	0,55	0,50	0,50	0,45	0,45	0,55	0,55	0,45	0,55	0,55	0,50	0,45	0,55	0,50	0,45
Minimum strength class	C12/15	C20/25	C25/30	C30/37	C30/37	C30/37	C35/45	C35/45	C30/37	C30/37	C35/45	C30/37	C25/30	C30/37	C30/37	C30/37	C30/37	C35/45
Minimum cement content[c] (kg/m^3)	–	260	280	280	300	300	320	340	300	300	320	300	300	320	340	300	320	360
Minimum air content (%)	–	–	–	–	–	–	–	–	–	–	–	–	4,0[a]	4,0[a]	4,0[a]	–	–	–
Other requirements	–	–	–	–	–	–	–	–	–	–	–	Aggregate in accordance with EN 12620 with sufficient freeze/thaw resistance				–	Sulfate-resisting cement[b]	

[a] Where the concrete is not air entrained, the performance of concrete should be tested according to an appropriate test method in comparison with a concrete for which freeze/thaw resistance for the relevant exposure class is proven.

[b] Where sulfate in the environment leads to exposure classes XA2 and XA3, it is essential to use sulfate-resisting cement conforming to EN 197-1 or complementary national standards.

[c] Where the *k*-value concept is applied the maximum *w/c* ratio and the minimum cement content are modified in accordance with 5.2.5.2.

Questions to practice

1. What factors should be considered when designing concrete for a foundation on a site previously used as agricultural land? Justify your answer.
2. Explain how steel reinforcement in concrete is electrochemically protected against corrosion.
3. As part of a structural health monitoring team assessing a concrete girder of a motorway bridge, you measure an average carbonation depth of 20 mm after 16 years. If the clear cover to the reinforcement is 40 mWm, estimate the time it will take for the carbonation front to reach the reinforcement.
4. Justify the key mix design parameters for concrete intended to resist carbonation-induced corrosion of reinforcement.
5. Critically analyse why a low aggregate-to-cement ratio is preferable when designing fire-resistant concrete.
6. Explain why repeated wetting and drying conditions in a structural reinforced concrete member increase the risk of corrosion, providing justification based on the underlying mechanisms.
7. Briefly explain two key chemical ingredients and their sources that trigger ASR in concrete and its controlling measures.

Monower Sadique
ISBN 978-1-83608-477-8
https://doi.org/10.1108/978-1-83608-474-720251005

Chapter 5
Low carbon cement, SCMs, and embodied carbon of concrete

After completing this chapter, you should be able to

- understand the concept of low carbon cement and its significance in sustainable construction
- define supplementary cementitious materials (SCMs) and their role in reducing the carbon footprint of concrete
- explore the hydraulic and pozzolanic properties of materials and how they contribute to concrete performance
- identify common challenges associated with SCMs in concrete
- analyse how the efficiency factor of SCMs is incorporated into concrete mix design
- understand the confirmatory criteria for hydraulic cement as outlined in relevant standards
- learn how the strength of cementitious binders is measured, reported, and utilised for quality control
- explore the concepts of water demand and standard consistency in cementitious binders and their implications for mix design
- understand the setting time of cementitious binders, their measurement methods, and their impact on concrete performance
- analyse how pozzolanicity is measured and its importance in evaluating binder performance
- understand the physical and chemical properties of binders and their influences on concrete
- explore the basic physicochemical properties of traditional cement and commonly used SCMs, including fly ash, GGBS, and silica fume
- review and summarise the latest developments and proven benefits of using fly ash, GGBS, and silica fume in concrete to improve durability, sustainability, and performance
- examine the principles that define lower-carbon concrete and the strategies to achieve it
- explore how the introduction of Flex 350 will pave the way for a more sustainable concrete industry
- understand how the embodied carbon of concrete is calculated and reported in sustainable construction practices.

Introduction

Concrete accounts for 25% of the embodied carbon of construction in the UK, while cement which is vital for concrete accounts for up to 90% of the greenhouse gas (GHG) emissions associated with the production of concrete (Low Carbon Concrete Group, 2022). Therefore, various initiatives for reducing the carbon intensity of concrete through developing low-carbon cementitious binders and low-carbon concrete are gaining interest and have been progressively adopted by codes of practice globally. In doing this, the use of SCMs plays a crucial role in reducing the dependency on traditional Portland clinker. Therefore, understanding the physicochemical properties of Portland clinker as well as the potential SCM is essential. The use of SCM in concrete is not only limited to reducing the content of clinker but has also proved to enhance durability and appearance and reduce cost.

There has been significant research into the utilisation of silica fume, fly ash, and GGBS as SCMs and high utilisation figures have been reported, such as blast furnace cement (CEM III) in the British Standards. However, there are also barriers to using SCM in concrete and standards have specified various confirmatory criteria for the physical and chemical properties of cementitious binders to be used in concrete. The proportional reduction of the strength of composite cement when compared with that of control has been reported in various studies, especially at early ages. To address this, modified design principles are also suggested. Different types of SCMs other than the standard listed materials such as calcined clay, sewage sludge ash, rice husk ash, wood ash, sugar cane bagasse ash, corn cob ash, kaolin waste, perlite, diatomite, air pollution control residue, and so on have been studied as SCM. Much of the research has been concluded by activating the SCMs with cement. At the same time, a consistent and comparable approach to reporting the embodied carbon of cementitious materials and lower-carbon concrete is essential to gauge the achievement of net zero concrete.

Low carbon cement

Although the technology has changed greatly, Portland cement technology still relies on the same basic raw materials and on volcanic heat (around 1450°C). As discussed in Chapter 1, 26 out of 27 family members of common cement (CEM II to CEM V) contain different proportions of pozzolanic and/or latent hydraulic additions or SCM. Jones (2019) reported significantly lower emission factors associated with PFA (0.004 $kgCO_2/kg$) and GGBS (0.0416 $kgCO_2/kg$) compared to cement clinker (0.97 $kgCO_2/kg$), justifying their application in low carbon concrete. The consumption of PFA, GGBS, and SF indicates that globally significant advances in using these materials as SCM have been achieved through extensive research. Global SF consumption of around 300 000 Mt per annum indicates its wide utilisation as SCM (Dunster, 2009). Out of an annual three million tonnes of blast furnace slag (BFS) produced in the UK by integrated steelmaking facilities, 75% is converted into GGBS and used by concrete manufacturers (Waste & Resources Action Programme, 2007). At the same time, around 50% of PFA sold was utilised as a cement replacement in cementitious products and as structural fill in 2009 (UKQAA, 2010). A general-purpose cement that contains additions is termed low carbon cement – for example, based on the ICE database (Jones, 2019), blast furnace cement CEM III/C-S contains up to 95% GGBS and has 83% lower embodied carbon than Portland cement CEM I, which has up to 95% clinker. The embodied carbon (emission factor) of some general-purpose cement and its alternatives specified in BS EN 197 (BSI, 2019a) are illustrated in Figure 5.1. It is evident from this figure that the greater the non-clinker content (of SCM), the lower the embodied carbon in the cement, which is also linked to the type of SCM. In addition to environmental, there are economic, technical, and durability benefits of using SCM, which are discussed later in this chapter.

Figure 5.1 Emission factor for different low carbon cement alternatives specified in BS EN 197 (BSI, 2019a) based on the ICE database (Jones, 2019) (author's own)

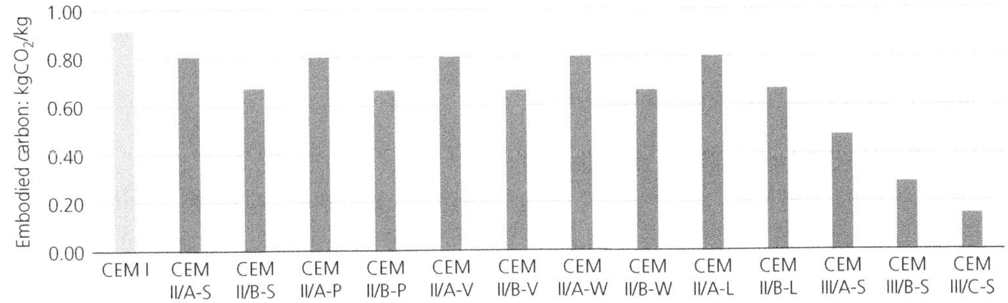

Supplementary cementitious materials (SCM)

The term 'additions' or SCM is defined in BS EN 206 (BSI, 2021c) as finely divided inorganic constituents used in concrete for improving certain properties or to achieve special properties. SCMs may be produced from natural sources (such as volcanic ash, kaolin clay, etc.) with minimal processing or may arise from wastes or by-products from other industries (such as GGBS, SF, PFA, etc.). One of the most significant current discussions in the construction sector is the reduction of carbon emissions from its production and application process. Research, development, and application of the SCM originating from different agricultural and industrial processes (as shown in Figure 5.2) play a key role in achieving this objective. The consumption of PFA, GGBS, and SF that contributes to circularity indicates that globally significant advances in using these materials

Figure 5.2 Different types of conventional and potential SCM sourced from different industrial and agricultural wastes and by-products (author's own)

as SCM have been achieved through extensive research. As discussed earlier, once a cementitious material contains addition/s, the term 'cement' should be replaced with 'binder', as additions also contribute to the hydration reaction, as shown in the following equation. In low carbon concrete, water/cement ratio and cement content specified by the relevant standard should be replaced by water/binder ratio and binder content, respectively.

$$\text{Binder} = \text{cement (CEM I or CEM II/A)} + \text{SCM} \tag{5.1}$$

Hydraulic and pozzolanic materials

Hydraulic and pozzolanic are two generic categories of conventional SCM. Hydraulic materials such as cement react directly with water to form cementitious compounds, while pozzolanic materials such as calcareous fly ash chemically react with $Ca(OH)_2$ (calcium hydroxide) in the presence of water to form compounds possessing cementitious properties. A pozzolan, either silicious or aluminous, that contains active silica possesses little or no cementitious properties in the absence of $Ca(OH)_2$ when mixed with water, unlike a hydraulic material that contains CaO (calcium oxide) and offers cementitious activity. Pozzolanic materials are classified based on their origin, either natural or artificial. Natural pozzolans are of volcanic origin or sedimentary rock such as volcanic ash which does not need any processing except grinding for their application as pozzolan. Artificial pozzolans such as fly ash, silica fume, metakaolin, and so on, which originally had no or only weak pozzolanic properties, need chemical and/or mechanical treatments such as calcination.

The properties of some generic SCMs specified in British Standards are schematically shown in Figure 5.3. From Figure 5.3 it can be stated that the transformation of any pozzolanic SCM into a hydraulic depends on a successful symbiotic combination between various SCMs. It is necessary to use a combination of oxide phases that are metastable in the presence of water and tend to convert to a more stable phase assemblage when hydrated, as reported by Gartner and Macphee (2011).

Problems and barriers in using SCM

In most cases, SCMs with lower reactivity are synthesised by conventional cement and, as a result, bulk replacement of the cement clinker is hindered (except blast furnace cement CEM III/C-S)

Figure 5.3 Schematic representation of pozzolanic and/or hydraulic properties of some generic SCMs (author's own)

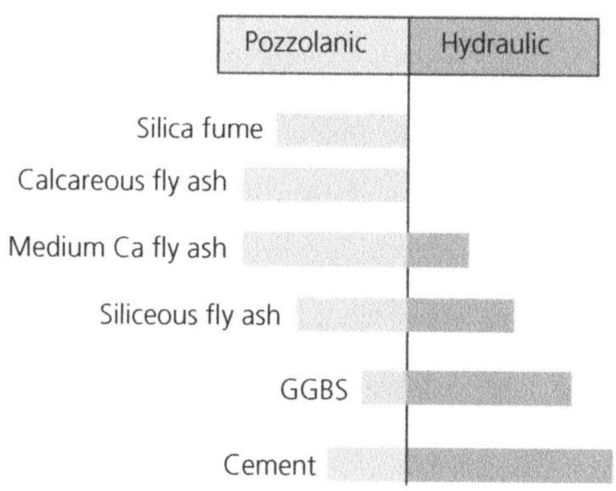

because neither of these materials is true pozzolana nor true hydraulic. However, in the presence of cement, they are found to be very reactive during the hydration reaction, where cement clinker acts as a catalyst. Additionally, processing potential SCMs through various activation techniques, such as sintering, chemical activation, and grinding, can enhance their reactivity. These methods allow for higher levels of SCM incorporation, leading to improved durability and strength properties in the final concrete product. Low carbon cement containing SCM requires testing to demonstrate that the concrete satisfies the performance requirements of the intended application considering the exposure environment and durability. Requirements for high early strength for striking formwork often give rise to a cement content and its reactivity that delivers in-service concrete strengths that exceed the specified strength as used in the design calculations. For instance, low carbon cement containing a high volume of SCM, such as CEM III/C-S, may have limited applications due to its slower setting time. It could take additional time (e.g. 56 days) to achieve the traditional 28-day design strength. In such cases, a collaborative effort between the designer, concrete supplier, and concrete technologist is essential with an aspect of flexibility in the specification to overcome the challenges in the effective application of low carbon concrete.

SCM efficiency factor

The efficiency or principles of the equivalent concrete performance concept or k-value concept suggested in BS EN 206 (BSI, 2021c) is a prescriptive concept. It is based on the comparison of the durability or strength performance of a reference concrete with cement ‘R’ against a low carbon concrete in which part of cement ‘R’ is replaced by SCM as a function of the water/binder ratio and addition content. This concept considers the activity of inorganic additions. The k-value is part of the additions in a low carbon concrete, which can be considered equivalent to cement or CEM I, having similar properties to the control (reference) concrete without additions. In this concept the term ‘water/cement’ ratio is replaced by ‘water/(cement + addition)’, and the (cement + k × addition) should not be less than the cement content specified for any exposure class. The k-value for CEM I or CEM II/A is 1. For conventional SCMs, the k-values specified in BS EN 206 (BSI, 2021c) are 0.4 for fly ash, 2.0 for silica fume, and 0.6 for GGBS – for example, in a low carbon fly ash concrete with effective cementitious or binder content of 500 kg/m^3 (assuming a fly ash ‘efficiency’ of 0.4 in accordance with BS EN 206 (BSI, 2021c)), if the proportion of fly ash, in relation to the total cementitious content (i.e. C + FA), is 40%, the amount of fly ash in the concrete mix can be determined as follows.

As 40% of the reference cement (C) has been replaced by fly ash (FA)

$$\text{so, } FA / (C + FA) = 0.4$$

$$\text{or } FA = 0.4\,C + 0.4\,FA$$

$$\text{or } 0.6\,FA = 0.4\,C$$

$$\text{or } C = 1.5\,FA$$

For effective cementitious content of 500 kg/m^3 and k-value of 0.4

$$C + (0.4 \times FA) = 500$$

$$\text{so, } 1.5\,FA + 0.4\,FA = 500$$

or $FA = 263\,kg/m^3$

and $C = 1.5 \times 263 = 395\,kg/m^3$

Conformity requirement of hydraulic cement in British Standards

There are 27 products specified under five main cement types in BS EN 197-1 (BSI, 2019a), which specifies the composition, specifications, and conformity criteria for these cement types. All 27 types of distinct cement products must meet defined mechanical, chemical, and physical properties. Different parts of BS EN 196 (BSI, 2011a, 2016a, 2020) specify the methods of determining these properties. The mechanical requirement includes the determination of standard mortar strength, and the physical requirement comprises the initial setting time and soundness. The chemical requirement involves the measurement of various chemical properties of the cementitious material. A summary of these properties along with their required values specified by BS EN 197-1 (BSI, 2019a) is tabulated in Table 5.1.

Measuring the strength of cementitious binder

According to BS EN 196-1 (BSI, 2020), strength conformity for any cement or cementitious binder needs to be conducted using a 40 × 40 × 160 mm prism mortar using the mould shown in Figure 5.4. The proportions in mortar by mass need to be one part binder, three parts CEN Standard sand, and one-half part water (water/binder ratio 0.50). The specimen should be cured for underwater curing and then strength tests carried out at different ages (2 days, 7 days, and 28 days). Each prism needs to be tested for flexural strength (3-point loading) and then each prism half tested for compressive strength by loading its side faces. The entire process is illustrated in Figure 5.5.

Table 5.1 Chemical requirements for cement (BSI, 2019a)

1 Property	2 Test reference	3 Cement type	4 Strength class	5 Requirement
Loss on ignition	BS EN 196-2 (BSI, 2005b)	CEM I CEM III	all	≤5.0%
Insoluble residue	BS EN 196-2 (BSI, 2005b)	CEM I CEM III	all	≤5.0%
Sulphate content (as SO_3)	BS EN 196-2 (BSI, 2005b)	CEM I CEM II CEM IV CEM V	32.5 N 32.5 R 42.5 N	≤3.5%
			42.5 R 52.5 N 52.5 R	≤4.0%
		CEM III	all	
Chloride content	BS EN 196-2 (BSI, 2005b)	all	all	≤0.1%
Pozzolanicity	BS EN 196-5 (BSI, 2011a)	CEM IV	all	Satisfies the test

Figure 5.4 Typical mould for preparing laboratory mortar specimens for testing binder strength (BSI, 2020)

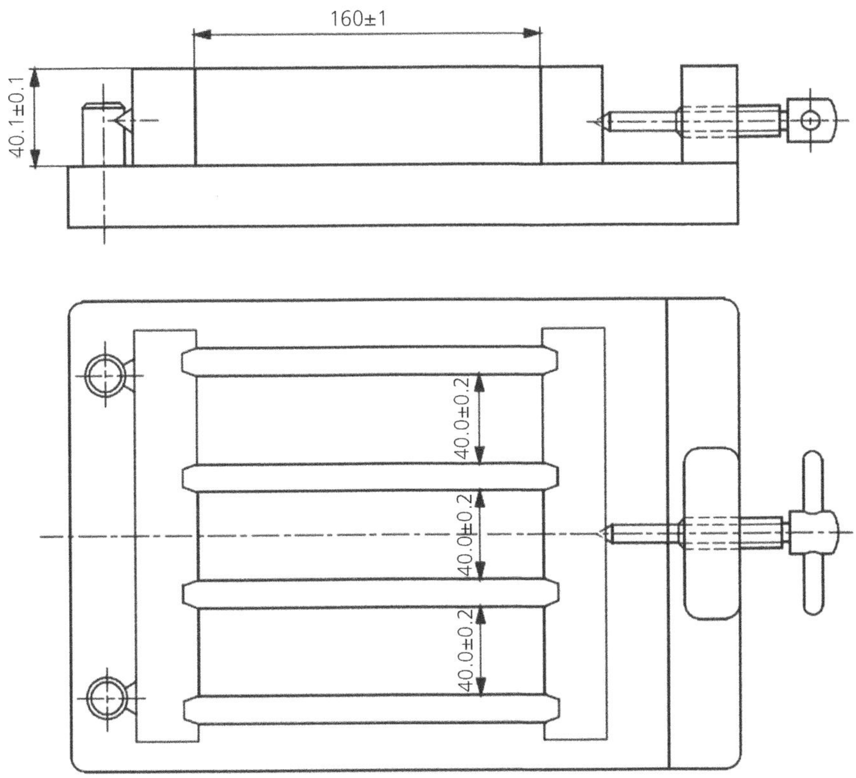

Consistency and water demand of cementitious binder

Water plays a crucial role in concrete or mortar not only by initiating and sustaining the hydration reaction through the dissolution of mineral phases and binding the network but also by enhancing the rheological properties. Water provides workability and consistency, which ensures proper packing and density by coating the particles during construction application. However, any additional quantity of water required for these two functions (hydration reaction and rheology) implies a concrete or mortar with a porous structure that essentially degrades the quality and durability. Therefore, understanding the amount of water needed for durable concrete with the required consistency for given constituents is essential.

The amount of water or water demand of mortar prepared with any cementitious and/or pozzolanic binder is dependent on the shape, surface morphology, inner structure of the particles, PSD, surface area, and reactivity of the binder. The 'water demand' is generally considered to be the quantity of water required for the preparation of a cement paste with standard consistency, as specified in BS EN 196-3 (BSI, 2016a).

Reduced water demand for grinding cement clinker and PFA (20%) mixture in a vibro mill for 3 min was reported by Sekulic *et al.* (1999). In the case of PFA, the creation of higher surface area and breakage of initial spherical particles on grinding was reported to show more water demand

Figure 5.5 Testing cementitious mortar sample for flexural and compressive strength (author's own)

(Felekoglu *et al.*, 2006). Moreover, as claimed by Aydin *et al.* (2010), the grinding operation breaks up large plerospheres (i.e. decrease of porous particles), which decreases particle roughness and thus water demand is reduced.

Standard consistency and setting time

Any hydraulic or pozzolanic binder paste of standard consistency has a specified resistance to penetration by a standard 100 mm plunger of a Vicat apparatus, as shown in Figure 5.6. Standard consistency is the water content required by the paste to allow the plunger to penetrate 6 ($\pm$2) mm. The standard consistency of any hydraulic or pozzolanic binder indicates the water demand which is dependent on the many factors discussed above.

Once water is added to an anhydrous cementitious binder, a plastic workable paste is formed which stiffens and loses mobility or workability through setting and consequently hardens due to the instigation of instantaneous hydration reaction which is influenced by chemical reactivity and phase composition of the binder. Adding a controlled amount of gypsum with clinker as a set regulator and grinding aid is widely practised in cement manufacturing.

The setting of cementitious binder is initiated by forming C-S-H and ettringite (AF_t) through the dissolution of aluminate (C_3A) and alite (C_3S) during hydration. A binder with a higher proportion of C_3A and C_3S is known to have a lower setting time. In cement chemistry, this phenomenon is described by the initial and final setting times of a binder, which are critical for determining the end of the workability phase and the beginning of the hardening process in concrete or mortar made with that binder. In the case of working in extreme weather conditions, such as very high temperatures in the Gulf region or freezing temperatures in the Nordic region, or when the concrete mixing

Figure 5.6 Vicat apparatus for determination of standard consistency and setting times. 1-mould, 2-weights, 3-base plate, 4-container, 5-water, 6-air vent ($\phi \approx 1.5$), 7-air vents, 8-needle for final setting time (BSI, 2016a)

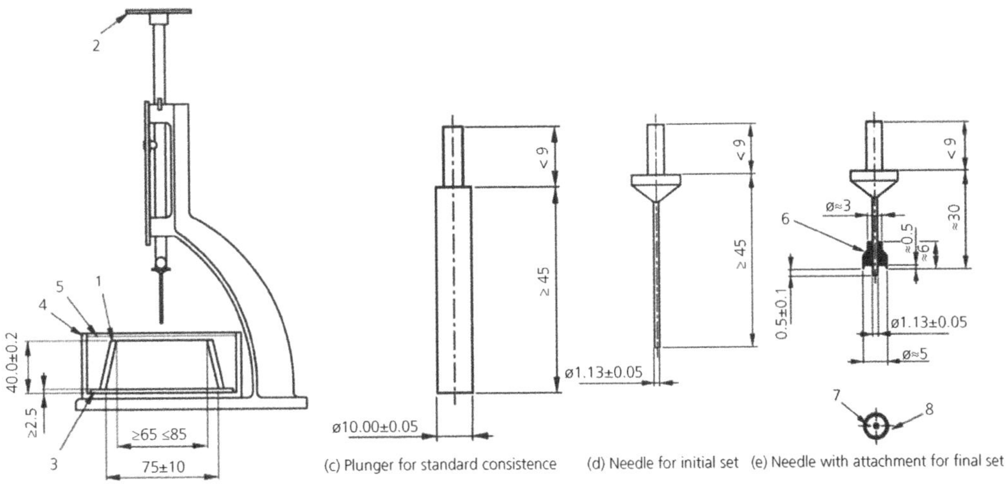

plant is far away from the project, or for a long concrete pour for mass concrete, the estimation of initial and final setting time is crucial. The time (in minutes after mixing water) required for a paste of a given binder to reach the initial set until mortar or concrete remains plastic and workable is called the initial setting time. The period (in minutes after mixing water) once the instigation of rigidity by the paste starts is called the final setting time. A Vicat apparatus is used to determine the setting times as specified by BS EN 196-3 (BSI, 2016a) where the paste with standard consistency needs to be used. Related to the strength class, BS EN 197 (BSI, 2019a) specifies a minimum characteristic value for initial setting time in minutes with no requirement for final setting time.

It is important to note that the setting time determined for a given binder in paste form is primarily used for quality control during manufacturing or to assess its expected behaviour. However, this setting time cannot be directly applied to concrete made with the same binder, as it is influenced by factors such as the water content of the mix. It is important to note that rapid setting and rapid strength gain or final set and early strength of concrete or mortar are not necessarily the same. Rapid setting is the rate at which the binder loses its plasticity, whereas rapid strength gain is the rate at which the concrete gains strength, such as compressive strength development after 2 days. Setting times are arbitrary terms used to describe the gradual rheological changes of cementitious binder due to hydration reaction.

Soundness of cementitious binder

The soundness of a cementitious binder is significantly affected by the presence of periclase (MgO) or free lime (CaO) which could appear from the raw meal used or from incomplete reaction due to unburned clinker as in the case of cement production. Free lime hydrates after cement is hardened and this creates expansion and cracking (as hydrates need more space than lime) and makes the binder unsound. The soundness is determined by observing the volume expansion of binder paste of standard consistency as indicated by the relative movement of two arms of the mould of the Le Chatelier apparatus, as shown in Figure 5.7. As specified in BS EN 196-3 (BSI, 2016a), the process involves measuring the expansion in mm created by heating the 24-hour-old paste in a water bath at

Figure 5.7 Le Chatelier apparatus for determination of the soundness of cements. Dimensions are in mm (BSI, 2016a)

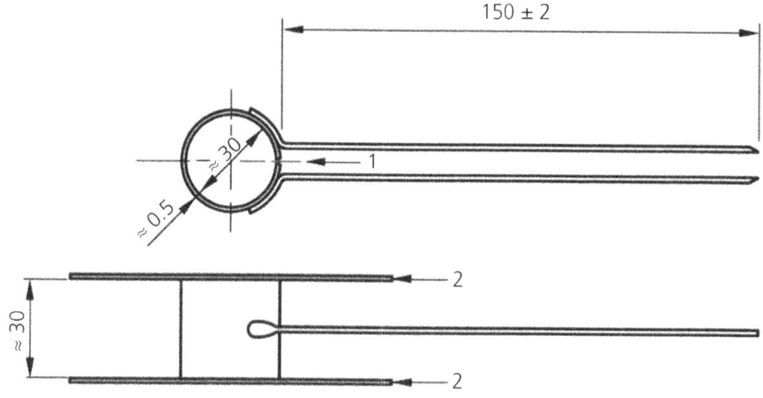

boiling temperature for three hours. BS EN 197 (BSI, 2019a) specifies a maximum value of 10 mm for cement with any strength class. A sound binder will eliminate the chance of volume change or cracking due to the creation of any delayed hydrates in hardened concrete.

Pozzolanicity of SCM

If a low carbon pozzolanic cement contains pozzolans (e.g. silica fume, fly ash, and pozzolana in CEM IV), the binder's pozzolanicity needs to be tested positive under the chemical requirement stated in BS EN 197 (BSI, 2019a). According to BS EN 196-5 (BSI, 2011a), the pozzolanicity of a pozzolanic binder is assessed by comparing the concentration of calcium ions present in the aqueous solution of cement after a fixed period (8 or 15 days), with the quantity of calcium ions capable of saturating a solution of the same alkalinity. The ion concentration is determined by titrating the previously titrated solution of the binder following the procedure and formulae stated in the standard. The OH and CaO ion concentration of any pozzolanic cement needs to be positioned below the saturation curve provided in BS EN 196-5 (BSI, 2011a) to satisfy the criteria for pozzolanicity, as shown in Figure 5.8.

Characterisation of physical and chemical properties of binders

Researchers around the world are continuously trying to invent novel cementitious materials to replace cement. In addition, to comply with strict environmental legislation, SCM is also used by cement manufacturers to improve the workability, durability, and strength of the products. Among the various materials, GGBS, SF, and PFA have been widely used by the global as well as the UK construction industries. The chemical and mineralogical composition of SCMs play an important role in the fresh and hardened and durability properties of low carbon concrete. Therefore, the development and production of new cementitious materials require advanced analytical characterisation to explain the relation between the physicochemical structure of the material and its properties. Highly integrated microelectronic structure analysis of surfaces with laser beams and X-ray fluorescence-aided devices are found to be helpful for analysing substances at micro-molecular level (Grasserbauer, 1989). In addition to characterisation, they provide important information, including interrelationships between physical, chemical, mechanical, and durability characteristics of the developed products.

Figure 5.8 Pozolanicity diagram for pozzolanic cement saturation curve (BSI, 2011a)

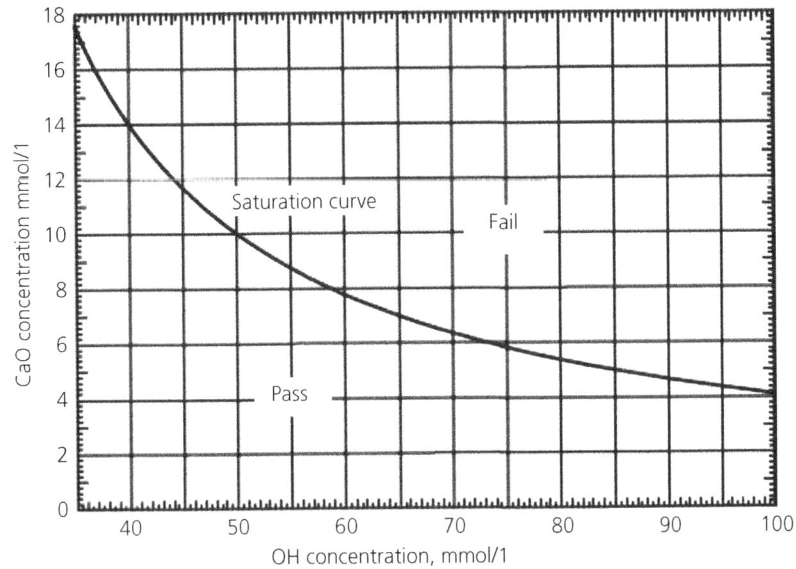

Two widely used techniques based on the characteristic behaviour of X-rays for material evaluation are energy dispersive X-ray fluorescence for elemental analysis and XRD for structural and phase composition studies. The effect of different particle shapes and sizes, as well as distributions of the parameters, play a significant role in the development and implementation of sustainable construction materials technology. The PSD not only influences the packing, but also the properties of fresh and hardened paste of the binder. Laser diffraction particle size analysis uses light scattering measurement across the widest dynamic range, from 40 nm to 2000 μm, in aqueous liquid mode. In the case of analysing the shape of materials and hydrates at micro-molecular level, SEM is used for high-resolution imaging of surfaces. Modern SEMs are usually equipped with an energy dispersive X-ray (EDX) detector which is able to analyse the elemental composition. The process of using SEM and EDX to simultaneously examine the morphology of cementitious binder and analyse its elemental composition by chemical microanalysis is called SEM/EDX (Hanke, 2001).

Basic physicochemical properties of cement

BS EN 197 (BSI, 2019a) specifies the chemical properties of cement as loss on ignition, insoluble residue, sulphate, and chloride content. The chemical and physical properties of cement CEM II/A are demonstrated in Tables 5.2 and 5.3. The mineralogical composition of typical CEM II/A cement is alite ($3CaOSiO_2$), belite ($2CaOSiO_2$), ferrite ($4CaOAl_2O_3{\cdot}Fe_2O_3$), calcite ($CaCO_3$), and periclase (MgO), as shown in Figure 5.9. The PSD shown in Figure 5.10 indicates that most of the particles are in the size range of 10–30 μm, whereas it contains a considerable amount of particles in the sub-micron region, thus creating a 13.4 μm mean diameter. The SEM view shown in Figure 5.11 reveals that the cement particles are composed of spherical and crystal-shaped particles which are also responsible for showing less water demand for standard consistency.

Table 5.2 Typical physical properties of conventional SCMs and cement

	CEM II/A	Fly ash	GGBS	SF
SSA (Blaine): m^2/g	3493	4092	4567	25140
Density: g/cm^3	3.205	2.220	2.750	2.39
Standard consistency: %	30	25	28	65
Activity index: %	–	106	–	–

Table 5.3 Typical elemental composition of conventional SCMs and cement

	CEM	Fly ash	GGBS	SF
SiO_2	19.60	51.20	30.00	97.00
Al_2O_3	4.71	24.34	10.00	0.5
Fe_2O_3	3.25	10.17	0.80	
CaO	64.00	2.79	35.00	
MgO	1.17	1.46	6.00	
Na_2O	0.27	1.28	0.20	1.0
K_2O	0.73	2.57	1.95	
SO_3	2.94	0.26	0.00	
TiO_2	0.26	1.01	0.02	
LOI	3.22	4.30	0.94	2.0

Figure 5.9 Typical powder XRD of CEM II/A, fly ash, and GGBS (Q-quartz, H-hatrurite, K-calcium aluminium magnesium silicate) (author's own)

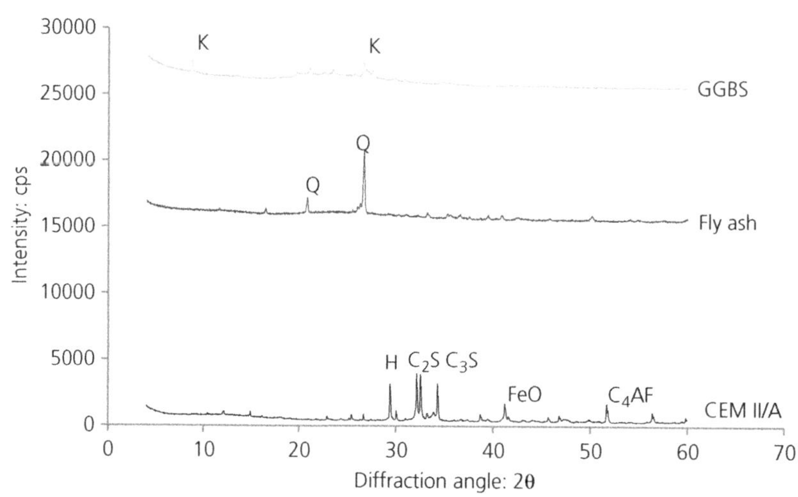

Figure 5.10 PSD of CEM II/A and common SCMs (fly ash, GGBS, silica fume) (author's own)

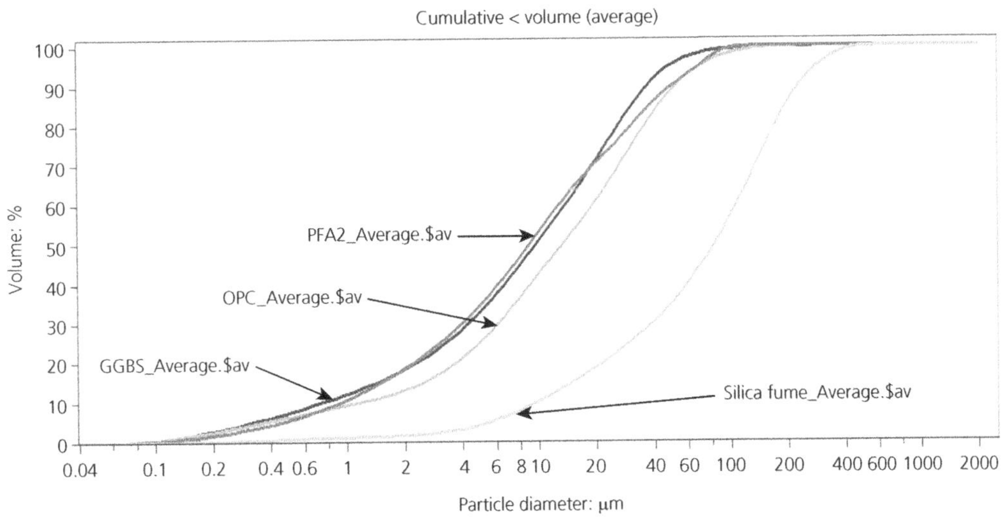

Figure 5.11 SEM image of dry CEM II powder (author's own)

Basic physicochemical properties of fly ash

Using natural coal as a fuel for energy generation results in huge amounts of waste fly ash and bottom ash. Fly ash or PFA is collected by electrostatic or mechanical precipitation during the de-dusting of the flue gases of furnaces before discharging to the atmosphere fired with pulverised coal. Depending on the type of coal and furnace of the boiler, the process usually generates spherical silicious or calcareous fly ashes. Fly ash to be used as SCM in concrete needs to be tested for various physical (e.g. activity index, fineness, water demand, setting time, etc.) and chemical (e.g. loss on ignition, the content of SO_3, CaO, SiO_2, etc) properties and the conformity requirement set out in table 2 of BS EN 450-1(BSI, 2012a). Activity index or strength ratio of given fly ash at 28 days

Figure 5.12 SEM images of typical fly ash particles at different magnifications (author's own)

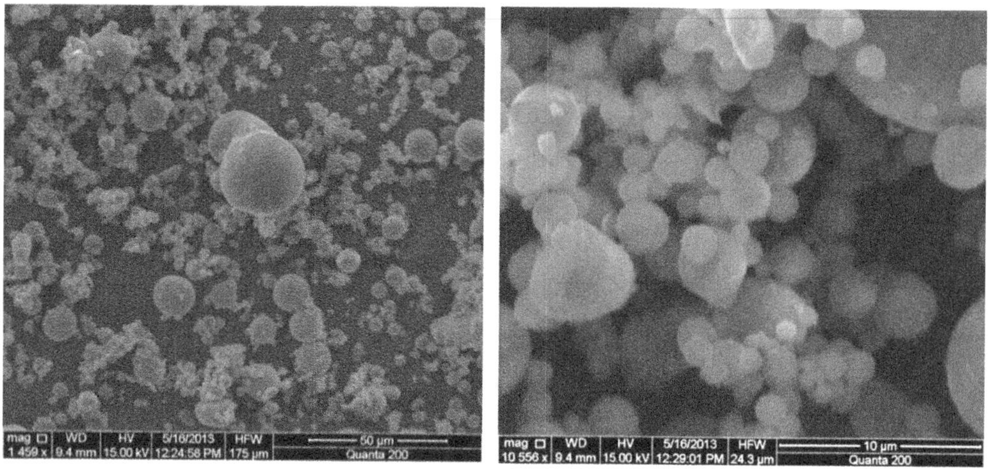

and at 90 days shall not be less than 75% and 85%, respectively. During the activity index test, the compressive strength of standard mortar bars prepared following the procedure specified in BS EN 196-1 (BSI, 2020) where 25% of reference cement is replaced with given fly ash is compared with standard mortar bars prepared with 100% reference cement, tested at the same age. Portland fly ash cement CEM II allows 6–35% clinker replacement by fly ash.

Siliceous fly ash which is pozzolanic in nature consists essentially of reactive silica (SiO_2) and alumina (Al_2O_3) where the content of reactive lime (CaO) is limited to 10% by mass. Calcareous fly ash, which has hydraulic and/or pozzolanic properties, consists of reactive CaO, Al_2O_3, and SiO_2 where the content of CaO shall not be less than 10%. All fly ashes at UK power stations are siliceous, although their fineness, colour, and reactivity vary significantly between stations. The disparity in terms of chemical composition, colour, fineness, particle shape, loss on ignition (LOI), the content of organic matter and water demand of fly ashes taken from two different power plants in the UK have been reported to result in approximately a 10% variation in activity index (Sadique and Coakley, 2016). As shown in Figure 5.9, the principal crystalline peaks identified in fly ash are quartz, with PSD similar to GGBS as shown in Figure 5.10. The spherical particles of fly ash as shown in Figure 5.12 justify its lower standard consistency and water demand.

Basic physicochemical properties of GGBS

GGBS is a vitrified glassy powder made by rapid cooling and grinding of slag smelt which is a by-product of the iron-making industry through smelting iron ore in a blast furnace. The main chemical constituents of GGBS are lime (CaO), silica (SiO_2), alumina (Al_2O_3), and periclase (MgO), as shown in Table 5.3; it has an amorphous nature with principal crystal peaks identified as calcium aluminium magnesium silicate, as shown in Figure 5.9. Due to differences in iron ore sources, the Al_2O_3 and MgO content of GGBS were found to vary significantly, while the levels of CaO and SiO_2 generally remain consistent. Its fineness is similar to cement and is off-white in colour, lighter than cement and fly ash. GGBS to be used as SCM in concrete needs to be tested for various physical (e.g. activity index, fineness, setting time, etc.) and chemical (e.g. loss on ignition, content of SO_3, MgO, etc.) properties and the conformity requirement set out in BS EN 15167-1

Figure 5.13 SEM image of GGBS powder (author's own)

(BSI, 2006). For the activity index of GGBS, the binder of the test mortar should contain 50% cement and 50% GGBS following the procedure specified in BS EN 196-1 (BSI, 2020). The activity index of a given GGBS at 7 days and 28 days shall be not less than 45% and 70%, respectively. GGBS possesses both latent pozzolanic and hydraulic properties. If a higher volume (as in the case of CEM III/C) of Portland clinker is replaced with GGBS then there will still be enough calcium hydroxide (C-H) compound to form the cementitious properties and the higher level of replacement will not create a substantial shortage of C-H. Therefore, the global standards specify a higher rate of Portland clinker replacement with GGBS compared to other SCMs. Blast furnace cement CEM III allows 35–95% clinker replaced with GGBS. Although concrete with a high GGBS content (> 50%) shows noticeably lower early strength at 7 days, it can achieve strength comparable to reference concrete (100% cement) after 56 days with 70% GGBS content (CSMA, 2024). The presence of irregular crystal-shaped particles in GGBS, as illustrated in Figure 5.13, resembles those found in cement. This similarity explains why GGBS exhibits a water demand comparable to that of cement and higher than that of fly ash, as detailed in Table 5.2.

Basic physicochemical properties of SF

Silica fume is a byproduct of silicon metal and ferro-silicon alloys in electric arc furnace consisting of extremely fine spherical particles smaller than 10^{-6} m containing at least 80% by mass of amorphous silica (SiO_2). SF is grey in colour and highly pozzolanic with no hydraulic properties. Due to its superfine particle size, it is often densified for transportation to increase its bulk density or converted into a slurry with water. SF particles often group and form agglomeration as shown in PSD and under petrographic analysis, as shown in Figure 5.10 and Figure 5.15, respectively. The physicochemical properties shown in Table 5.3 indicate that SF is composed of 97% silica. The XRD powder diffraction shown in Figure 5.14 indicates that SF contains amorphous silica particles with high SSA, thus increasing the water demand for standard consistency. The coarser PSD shown in Figure 5.10 is due to the agglomeration of superfine particles, which was also revealed in the SEM micrograph, as shown in Figure 5.15. Like fly ash and GGBS, SF to be used as SCM in concrete needs to be tested for various physical (e.g. activity index, fineness) and chemical (e.g. loss on ignition, SiO_2 content) properties and the conformity requirement set out in BS EN 13263-1

Figure 5.14 Diffractographs of amorphous silica fume particles (author's own)

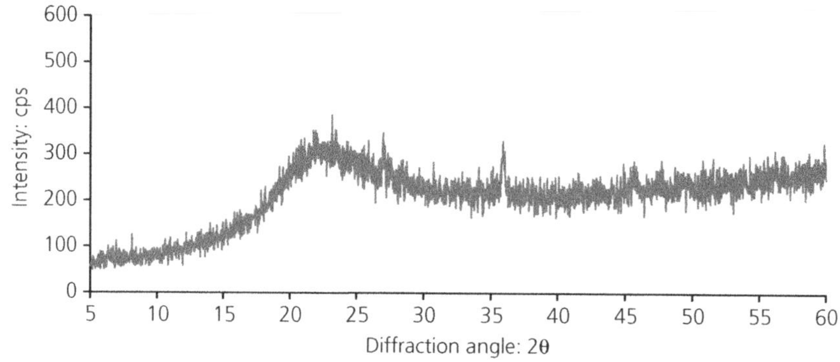

Figure 5.15 SEM image of superfine agglomerated condensed silica fume particles (author's own)

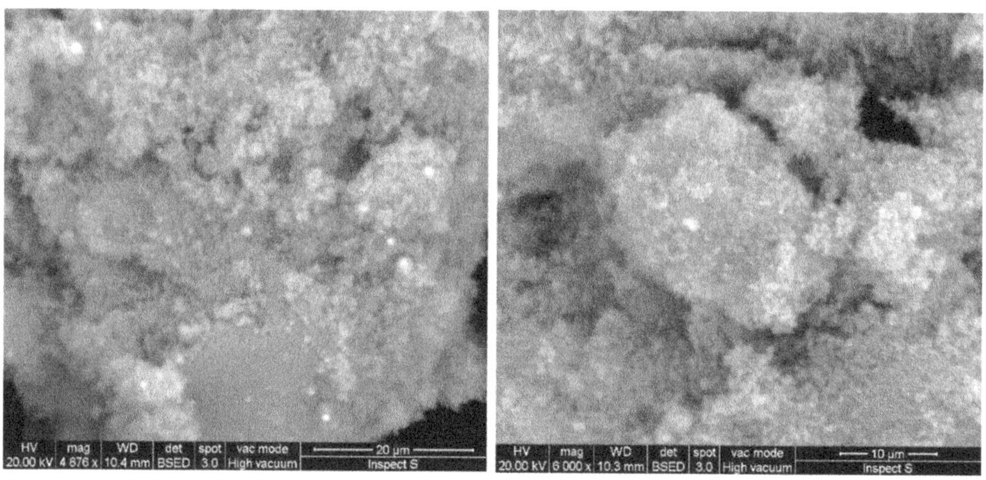

(BSI, 2009b). As specified in BS EN 13263 (BSI, 2009b), for the activity index of SF, the binder of the test mortar should contain 90% cement and 10% SF following the procedure specified in BS EN 196-1 (BSI, 2020). The activity index of a given SF shall be at least 100% when tested at a mortar age of 28 days. Due to its superfine nature, SF is the most reactive pozzolan of the conventional SCMs. Portland silica fume cement CEM II/A-D allows 6–10% replacement of clinker with pozzolanic SF.

A considerable amount of literature has been published on these conventional SCMs, and Tables 5.4 to 5.6 summarise these studies. The physical and chemical properties of cement clinker and traditional SCMs that influence concrete properties at fresh and hardened states are discussed in the following sections.

Benefits of fly ash concrete

As a purely pozzolanic mineral, the addition of fly ash in concrete is generally limited to low to medium levels. However, its significantly low emission factor, along with its physicochemical

Table 5.4 Key development of using PFA as SCM

Reference	Findings
Tan and Pu (1998)	20% of finely ground PFA or GGBS in 80% cement showed more strength than the control, whereas the combined ternary blend of PFA, GGBS, and cement (10:10:80) showed higher strength than the binary blends. The strength development in the ternary blend was similar to the binary blend of 6% incorporated SF in cement.
Li *et al.* (2000)	The hydration and activation of PFA were very low in the PFA–cement system at room temperature. But at elevated curing temperature or when fly ash was activated by blended alkali (sodium hydroxide and calcium hydroxide) and gypsum, not only was the hydration of fly ash enhanced, but also more rational and relatively stable hydrates were formed.
Arjunan *et al.* (2001)	Effects of chemical and mechanical activation on low calcium PFA were studied in an 80% cement–20% PFA system. Various chemical activators applied were hydroxides of calcium and sodium in different concentrations: Na_2CO_3, Na_2SO_4, NaCl, NH_4NO_3, $NaNO_3$, $Ca(NO_3)_2$, $CaCl_2$, including binary and ternary mixtures of some of the activators. Among all activators, only NaOH in low concentration showed the highest performance. Combined mechanical and chemical activation showed better performance than untreated PFA at all ages.
Chindaprasirt *et al.* (2007)	The resistance of concrete to chloride penetration increases with incorporation of fly ash and with an increase in fly ash fineness.
Yazici (2008)	High-performance high-volume fly ash self-compacting concrete (SCC) was developed in this study. 60% PFA replacement in the PFA–cement system caused a reduction in both early and ultimate compressive strength, whereas the inclusion of 10% SF in the ternary system positively affected compressive strength and exceeded the control mixture at 28 days using superplasticiser.
Winnefeld *et al.* (2010)	In high calcium fly ashes, fewer alkali aluminate silicate hydrates and a much more porous microstructure are formed compared to the low calcium fly ash. Thus alkaline activation does not seem to be a suitable alternative to use in binary and ternary systems with GGBS and cement.
Müllauer *et al.* (2015)	The leaching behaviour of major and trace elements from cement–fly ash and cement–GGBS concrete was investigated, focusing on cement replacement levels of up to 40% with fly ash and 70% with GGBS, at equivalent water/binder ratios of 0.4 and 0.6. It was observed that a lower water/binder ratio reduced alkali release due to decreased porosity, despite fly ash containing higher alkali levels than cement. The total vanadium (V) content in concrete significantly increased with higher fly ash dosages; however, only about 2% of the total V was available for release, attributed to the high pH of the eluate. Chromium (Cr) release was not directly proportional to the replacement levels of fly ash or GGBS, even though Cr content increased with higher dosages of these materials. A minimal dissolution of Cu, Ni, Co, and Pb was found for their low solubility.
Zhang *et al.* (2024)	Fly ash cenospheres were used to develop ultra-light concrete with a dry density of 626–682 kg/m^3 avoiding using any air-entraining agents or foam. A strong bond between cenospheres and cement matrix generated around 5.9 MPa, attributed to the formation of ettringite, C-H, and C-(A)-S-H through pozzolanic reaction.

Table 5.5 Key developments using GGBS as SCM

Reference	Findings
Osborne (1999)	Technical benefits such as reduced heat evolution, lower permeability, higher strength at later ages, decreased chloride ion penetration, increased resistance to sulphate attack, and alkali–silica reaction were affirmed with 70% cement replacement. However, a requirement for good early curing for controlling the adverse effects of higher rates of carbonation, surface scaling, and frost attack has been suggested for blast furnace cement.
Babu and Kumar (2000)	Efficiency evaluation of pozzolans like fly ash and silica fume was also found to be appropriate for the evaluation of GGBS. Moreover, the overall strength efficiency was found to be a combination of a general efficiency factor, depending on age, and a percentage efficiency factor, depending on the percentage of replacement, as was the case with fly ash and silica fume.
Li and Zhao (2003)	The high calcium and magnesium content of GGBS accelerates the rate of hydration on incorporation in a binary system of fly ash and cement. The resulting ternary blend also shows superiority to both binary and control against H_2SO_4 attack.
O'Rourke *et al.* (2009)	Calcium sulphate-based binder (using anhydrite II as dominant) in a system of 7% cement:20% GGBS:70% gypsum:5% lime was found to achieve considerable strength but to be susceptible to moisture-induced deterioration due to the high solubility of gypsum. In a GGBS-based binder (containing a small amount of anhydrite II and gypsum) in a system of 27% cement:66% GGBS:4% gypsum:3% lime, however, not only was remarkable improvement in early strength development found, but also the strength and sulphate resistance at later ages remained unaffected.
Yang *et al.* (2012)	XRD peaks of calcium silicate hydrate (C-S-H) gels were more intensified in water curing than for air-dried curing of alkali-activated (AA) GGBS in a cement-free system. The growth and intensity of C-S-H gels were consistent with higher strength development, whereas the strength gain ratio at the early ages of water-cured AA-GGBS mortars was lower than that of the companion air-cured mortars.
Yu *et al.* (2015)	The accelerated reaction of GGBS with $Ca(OH)_2$, leading to the formation of gel, was noted to outperform the slower interaction between fly ash and portlandite. This behaviour was particularly significant in the development of UHSC using a binary system where cement was partially replaced with GGBS, fly ash, or limestone powder. The mix also incorporated nano-silica, microsand, and superplasticiser. The superior mechanical strength of UHSC containing GGBS, observed after 28 and 91 days, compared to mixes with fly ash or limestone powder, was attributed to the enhanced pozzolanic reaction of GGBS. This reaction benefited from the increasing availability of $Ca(OH)_2$ after the initial 5 days of curing, while the fly ash surface acted as a Ca^{2+} sink through chemisorption, influencing hydration dynamics.
Blotevogel *et al.* (2024)	The reactivity of GGBS is significantly influenced by its elemental composition, particularly the levels of CaO, SiO_2, Al_2O_3, and MgO, as well as the activation techniques employed, including cement-based, alkali, and supersulfated systems. Higher CaO and MgO content consistently enhance reactivity across all ages, regardless of the activation system. In contrast, increased Al_2O_3 content primarily boosts reactivity during early ages. Optimal activation is closely tied to the specific elemental composition of GGBS, underscoring the need for customised activation systems based on the elemental composition of GGBS for industrial-scale applications. For instance, artificially increasing the Al_2O_3 content in slags could be used to achieve higher early strength in concrete products.

Table 5.6 Key developments using silica fume as SCM

Reference	Findings
Li (2004)	Incorporation of 4% nano-SiO_2 (replacing cement) in a system of high-volume high-strength fly ash concrete (containing 50% PFA) leads to an increase of both short-term and long-term strength compared to high-volume high-strength fly ash concrete and control concrete.
Collepardi *et al.* (2004)	The binary blend of concrete containing 13% SF was found to perform better in terms of strength and durability than the binary blend of PFA at both normal (20°C) and elevated (65°C) curing.
Heikal *et al.* (2006)	Compressive strength values of control pastes are higher than cement + 10 wt.% SF pastes in the absence of polycarboxylate SP, whereas in the presence of 1.5 wt.% SP, the compressive strength values of cement + 10 wt.% SF were found to be higher than those of control pastes.
Radlinski and Olek (2012)	A synergistic effect for combined physical (particle packing) and chemical (increased hydration) effects was achieved by incorporating silica fume in a fly ash–cement system.
Okoye *et al.* (2016)	Fly ash-based geopolymer concrete activated with NaOH and Na_2SiO_3 demonstrated higher compressive, tensile, and flexural strengths compared to the control mix, attributed to increased compactness and a denser microstructure. However, this improvement was accompanied by reduced workability. Notably, when 40% silica fume was incorporated, the strength increased significantly, by approximately 100% after 3 days and 50% after 28 days. This enhancement was attributed to the presence of silicate ions in the alkaline solution, which contributed to accelerated geopolymerisation.
Ahmad *et al.* (2022)	The addition of 5% silica fume to a cement–natural pozzolana system, where 20% of the cement was replaced with NP, not only offset the early-age compressive strength loss but also resulted in a 9% higher strength compared to the control after 360 days. Furthermore, the inclusion of silica fume significantly reduced drying shrinkage and water penetration depth, while enhancing resistance to reinforcement corrosion, sulphate attack, and salt weathering, demonstrating its efficacy in improving durability and long-term performance relative to the control mix.

properties that provide notable improvements in the physical, mechanical, and durability aspects of concrete in both fresh and hardened states, have achieved widespread attention in the construction industry. The strength of concrete containing fly ash largely relies on the reactivity and fineness of fly ash in addition to the binder content and mix design. The spherical, smooth faces, fine particles, and lower water demand (standard consistency lower than cement) of fly ash result in improved rheology, reduced bleeding, and help to reduce the water content for a given workability, making the concrete more thixotropic with a higher compressive strength. While standards permit up to 35% clinker replacement with fly ash, researchers and manufacturers worldwide are striving to increase this replacement level in high-volume fly ash (HVFA) concrete without compromising strength. In cases where strength is adversely affected, these mixes are being redirected toward non-structural applications, such as road sub-bases, trench fills, and soil stabilisation. One of the challenges for using HVFA in concrete is low early strength, as concrete early strength largely depends on the content of clinker and the reactivity of the ashes. To counteract the reduction in early strength due to HVFA, mechano-chemical activation techniques, such as ternary or quaternary blends and grinding, have been effectively applied.

Mechanical activation of fly ash by grinding breaks the agglomeration of glass particles and enhances the activity. Ultrafine fly ash with particle sizes ranging from 0.5–5 μm was obtained by grinding and separation was found to offer 24.27% more compressive strength than its unground counterpart where 80% cement was replaced (Cai *et al.*, 2024). Enhanced particle packing and reactivity in ultrafine fly ash reduce mix porosity and water demand, contributing to improved strength.

Remarkable improvements were observed, including early strength gains after 2 days and a 28-day strength reaching 70 MPa, exceeding control (100% cement) by replacing 70% of the cement with fly ash (45%), cement kiln dust (CKD) (6.5%), and GGBS (18.5%) (Sadique and Coakley, 2016). However, this study used a high binder content of 580 kg/m^3 for structural grade HVFA concrete, achieving a moderate early strength of around 20 MPa compared to the control concrete's 37.5 MPa. Adopting a design principle that specifies 56-day strength, rather than the conventional 28-day benchmark, could provide a more sustainable approach for low-carbon concrete, allowing for greater clinker replacement at lower binder content without compromising structural integrity.

Benefits of GGBS concrete

GGBS is the most widely known alumina–silicate source used as a raw material in geopolymer/alkali-activated cement (AAC). Extensive laboratory studies have used GGBS in AAC formation, in addition to its established performance in Portland-slag cement, blast furnace cement, and composite cement.

As mentioned before, the high glass content resulting in rapid slag-smelt cooling makes the GGBS reactive. The latent hydraulic property and neutral water demand characteristics of GGBS allow high-volume cement replacement in concrete which is easy to compact with a desirable surface finish. Concrete with 60% GGBS has been shown to have reduced porosity and increased resistance to reinforcement corrosion, attributed to the formation of a greater amount of C-S-H (Cheng *et al.*, 2005), enhancing its durability in the marine environment. Similarly, no evidence of the thaumasite form of sulphate attack has been reported in concrete containing 70% GGBS when limestone aggregate is used (Higgins and Crammond, 2003).

Like fly ash, a mix where the clinker replacement level is over 50% with GGBS in either AAC or Portland-based cement, an increase in total binder content (kg/m^3) is required, which reduces the overall carbon savings potential – for example, a C32/40 class may require 420 kg of GGBS in AAC, whereas a Portland cement-based system may require 200 kg (Low Carbon Concrete Group, 2022). Annual global production of GGBS is about 10% of global cement use. However, the steel manufacturing industry is moving away from blast furnace technology to improve its carbon performance, phasing out blast furnaces, which will significantly impact GGBS's future supply. Therefore, mix design optimisation to balance the embodied carbon and material efficiency/availability has been suggested in the ICE *Low Carbon Concrete Routemap* (Low Carbon Concrete Group, 2022).

Benefits of SF concrete

The superfine nature of SF particles with extremely high surface area increases the water demand, which makes the concrete slightly stiff and cohesive and less susceptible to bleeding and segregation compared to normal concrete. However, for given workability, the use of a superplasticiser as an admixture in silica fume concrete is always suggested. Unlike GGBS and PFA, SF contains highly reactive amorphous silicon dioxide that enhances the dissolution of the C_3S component of cement, generating portlandite (C-H) and C-S-H gel compounds. This process not only densifies the concrete complemented by the filler effect of SF's superfine particles but also maintains early strength similar to reference concrete. In the case of the cement–silica fume system, the formation of densely compacted matrices at the ITZ is found to provide high strength as a consequence of improved bond (Toutanji and Elkorchi, 1995). As a result, SF provides both chemical and physical benefits, contributing to the creation of durable concrete with superior resistance to the ingress of harmful liquids or gases from the surrounding environment. Additional C-S-H gel was formed in a composite fibre-reinforced high-performance concrete where combined 5% fly ash and 5% silica fume replacing cement showed significantly improved acid resistance compared to concrete with 100% cement (Patil *et al.*, 2022).

Beyond mix design parameters, the strength of SF concrete is highly influenced by the content of SF, water/binder ratio, and the use of superplasticisers. A noticeable progressive increase in both compressive and flexural strength of concrete was reported (Wu *et al.*, 2016) for up to 20% silica fume replacement, with around 110 MPa compressive strength after 28 days. Figure 5.16 clearly illustrates the substantial impact of the water-to-binder ratio in determining the optimal dosage of SF in concrete at fixed superplasticiser content (Bhanja and Sengupta, 2003).

Figure 5.16 Influence of SF content and water/binder ratio on concrete strength (Bhanja and Sengupta, 2003). Courtesy of Elsevier

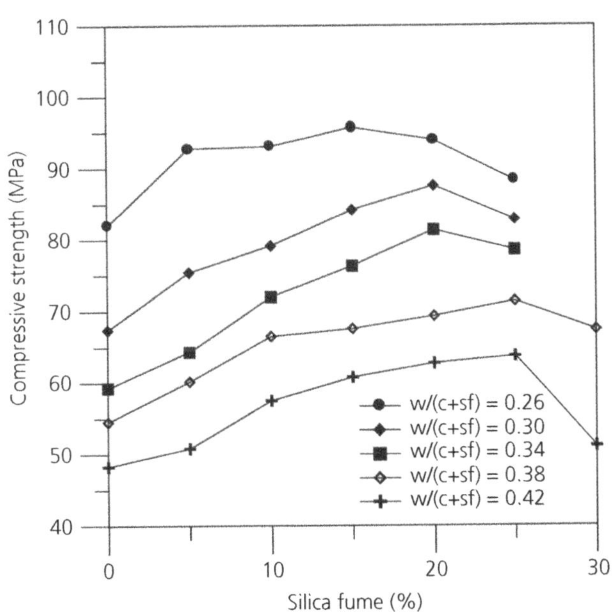

Durability and sustainability of concrete with SCM

The use of SCM immediately reduces the environmental burdens imposed by cement. In addition to the environmental benefits, the social, economic, and durability benefits arising from using SCM are driving their wider application; accordingly the *Low Carbon Concrete Routemap* endorsed by the UK Green Construction Board (Low Carbon Concrete Group, 2022) suggests increasing and optimising the use of SCMs with the adoption of additional multi-component binders into standards. The Burj Khalifa, Dubai, the world's tallest building, utilised a concrete mix incorporating a triple-blended binder of SF, low C_3A cement, and PFA to achieve an impressive compressive strength of 80 MPa (Elkem ASA, 2023). Given the harsh local conditions characterised by high chloride and ground with high sulphate levels, a highly durable and impermeable concrete with exceptional chloride resistance was developed. This innovative mix also set a world record by being pumped to a height of 606 metres against gravity. SCM was initially viewed as a cement replacement material, but currently, the most important reason for its use is the production of high-performance durable concrete, where binder with multi-component SCM provides enhancements in concrete properties.

For mass concreting or for thick structural elements when the heat of hydration is problematic, cement with low heat characteristics can be achieved by reducing the content of the more rapidly hydrating compounds in C_3S and C_3A by SCM. A 5500 m^3 continuous concrete pour, utilising 700 truckloads over 36 hours, was completed for the 3 metre-deep raft foundation of the Shard, the UK's tallest building (MPA The Concrete Centre, 2017). The C35/45 concrete mix, incorporating 70% GGBS, was specifically designed to control thermal cracking. Full strength was achieved in 56 days, showcasing the effectiveness of GGBS in controlling the heat of hydration in mass concrete while ensuring specified design strength development.

In modern architecture, exposed or visual concrete is intended to remain visible for the structure's entire lifespan, without the need for plastering or painting. Achieving the desired appearance, finish, and colour with consistent quality and durability can be effectively achieved using SCMs. In 2010, the Angel, a six-storey office building in London, was constructed using self-compacting exposed concrete with 36% fly ash (MPA The Concrete Centre, 2017). The spherical shape of the fly ash particles reduced water demand, eliminating the need for compaction while enhancing workability. This improved workability allowed the concrete to flow easily around complex interfaces of the in situ cast structural elements, preventing honeycombing and ensuring a smooth, high-quality finish.

Key design guidance for sustainable concrete

For low-carbon technology, several factors must be considered, including the safety of the design in terms of material selection and application, ease of construction, commercial feasibility, aesthetics, and sustainability. While BS EN 206 (BSI, 2021c) and BS 8500 (BSI, 2021a, 2021b) do not provide specific guidelines for sustainable concrete, MPA The Concrete Centre (2020) and the Low Carbon Concrete Group (2022) recommend the following key design principles for producing sustainable concrete.

- At the early stage of a project, a realistic balance of carbon reduction targets for concrete and the design, specification, construction method, and contract programme should be considered.
- Responsibly sourced concrete ingredients and reinforcement should be specified.
- When required and possible, 56 days of design strength rather than the conventional 28 days should be specified.

- Binders with a wider range of SCM specified in BS EN 197 (BSI, 2019a) should be selected, with the highest possible clinker replacement including those covered in BS EN 197-5 (BSI, 2021d) and BS 8500-2 (BSI, 2021b).
- The visual appearance and aesthetics of concrete should be considered when selecting SCMs.
- Use of admixtures should be promoted during material design.
- The use of recycled or secondary aggregates should not be over-specified and their local availability should be prioritised to avoid transport.
- During mix design, priority should be given to using the largest maximum aggregate size for achieving efficient placing and compaction.
- The best practices and new technologies in concrete mix design, batching, and production to manufacture low carbon concrete should be promoted sector-wide.

Lower carbon concrete

In October 2024, GCCA published a policy document (GCCA, 2024b) on numerical definitions in units of embodied carbon dioxide equivalent per cubic metre of product (ECO_2e/m^3), also referred to as GWP, for 'low carbon' and 'near zero' carbon emissions concrete products. Considering consistency and reliability, like the ICE *Low Carbon Concrete Routemap* (Low Carbon Concrete Group, 2022), GCCA also suggests that environmental product declarations (EPDs) are the preferred method for measuring and reporting the carbon footprint of concrete products.

BS EN 15804 (which replaced BS ISO 21930) specifies the core rules for a construction environmental product declaration (EPD) as a tool that provides quantified environmental information for a construction product or service on a harmonised and scientific basis (BSI, 2022). Embodied carbon dioxide equivalent ($kgCO_2e$) is defined as the total mass of GHG released during raw materials extraction, processing, transport, and manufacture converted to a carbon dioxide equivalent by multiplying the mass of each GHG by its corresponding GWP (cradle to gate: modules A1–A3 of BS EN 15804 (BSI, 2022)).

A lower carbon concrete must have at least 15% lower $kgCO_2e$ than the average for concretes in use across the market for the relevant strength class. To reduce the dependency on Portland clinker by using an alternative binder system (ABS) for lower carbon concrete, the British Standard Institution (BSI) has published BSI Flex 350 version 2 (BSI, 2024). However, BSI Flex is neither a standard nor a PAS, but it enables a standard to be rapidly developed to immediately fulfill stakeholders' demands. It recommends a framework for assessing lower carbon non-Portland-based concrete using ABS to facilitate their acceptance when designing and building structures. Depending on the physicochemical properties of the components of ABS, the binding mechanisms may involve hydration, carbonation, or alkali activation. ABS concrete should be specified as a designed concrete with compressive strength class, exposure class, setting time, and a target value for workability. The characteristic strength of 28 days or 56 days where suitable needs to be determined. Relative strength compared to 28 days for equivalent strength development of slow, normal, and rapid hardening ABS concrete needs to be determined, as shown in Figure 5.17 (Flex 350, BSI, 2024). Flex 350 is a performance-based guidance with responsibilities for the supply chain to support innovation towards lower carbon concrete.

Embodied carbon of concrete

The embodied energy of a construction product refers to the total energy required for its production and delivery, while embodied carbon (EC) quantifies the carbon emissions associated with that

Figure 5.17 Relative strength compared to 28 days for different strength development classes of ABS concrete (BSI, 2024)

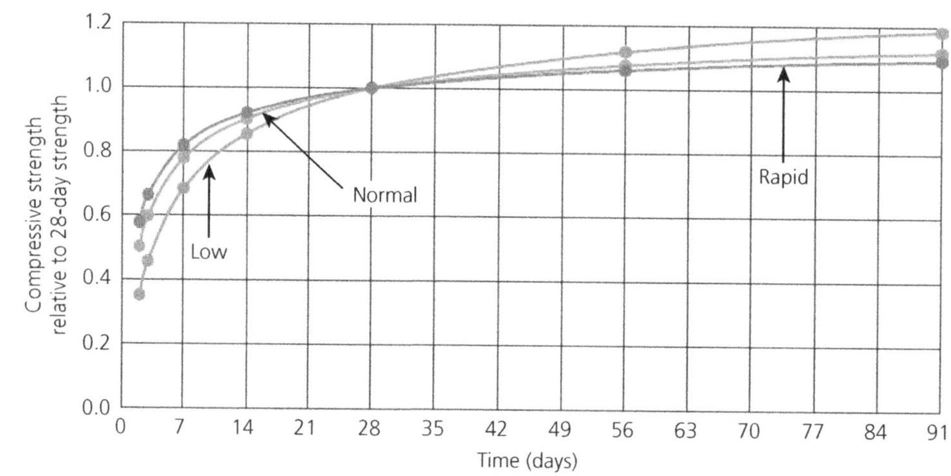

energy use. Additionally, energy is consumed during the operational phase of buildings or structures, known as operational energy, with its associated emissions termed operational carbon. In the built environment (buildings and infrastructure), energy and carbon are often used interchangeably as proxies, with energy referring to fuel from renewable or non-renewable sources and carbon representing the CO_2 emissions from fuel combustion. However, the carbon emitted per unit of energy varies depending on the fuel source. The Intergovernmental Panel on Climate Change has defined the GWP of each GHG that reflects its warming effect in comparison to CO_2; the carbon dioxide equivalent (CO_2e) refers to the mass of a given GHG multiplied by its GWP; and carbon factor refers to the kgCO_2e per unit of product. Since carbon emissions give a better indicator of the contribution of that energy to global warming and climate change, the ICE database (Jones, 2019) in the UK, like wider literature, relies only on EC and accordingly the database was developed for embodied carbon factors (kgCO_2e/kg) of commonly used construction materials (clinker, SCMs, steel, brick, etc.) to be used for calculating EC (kgCO_2e/m^3 of concrete or kgCO_2e/kg of concrete) of building products such as cast in situ or precast reinforced concrete.

$$\text{EC (kgCO}_2\text{e)} = \text{material quantity (kg)} \times \text{carbon factor (kgCO}_2\text{e/kg)} \quad (5.2)$$

The EPD serves as a standardised source for determining the EC of a building product, providing verifiable, accurate, and transparent environmental data. To ensure consistency and comparability in EPDs, BS EN 15804 (BSI, 2019d) establishes a framework and core rules for construction products, services, and processes, while BS EN 15978 (BSI, 2011b) outlines the calculation methodologies for the environmental performance of buildings. BS EN 15978 (BSI, 2011b) provides modular information for different lifecycle stages of a building's assessment while BS EN 15804 (BSI, 2022) suggests types of EPD concerning lifecycle stages covered, as shown in Figure 5.18.

There are a total of five stages, namely product (A1–A3), construction process (A4–A5), use (B1–B7), end of life (C1–C4), and beyond system boundary stage (D). Unless exempted

Figure 5.18 Types of EPD with respect to lifecycle stages and modules for the construction works assessment based on BS EN 15978 and BS EN 15804 (BSI, 2011b, 2022) (author's own)

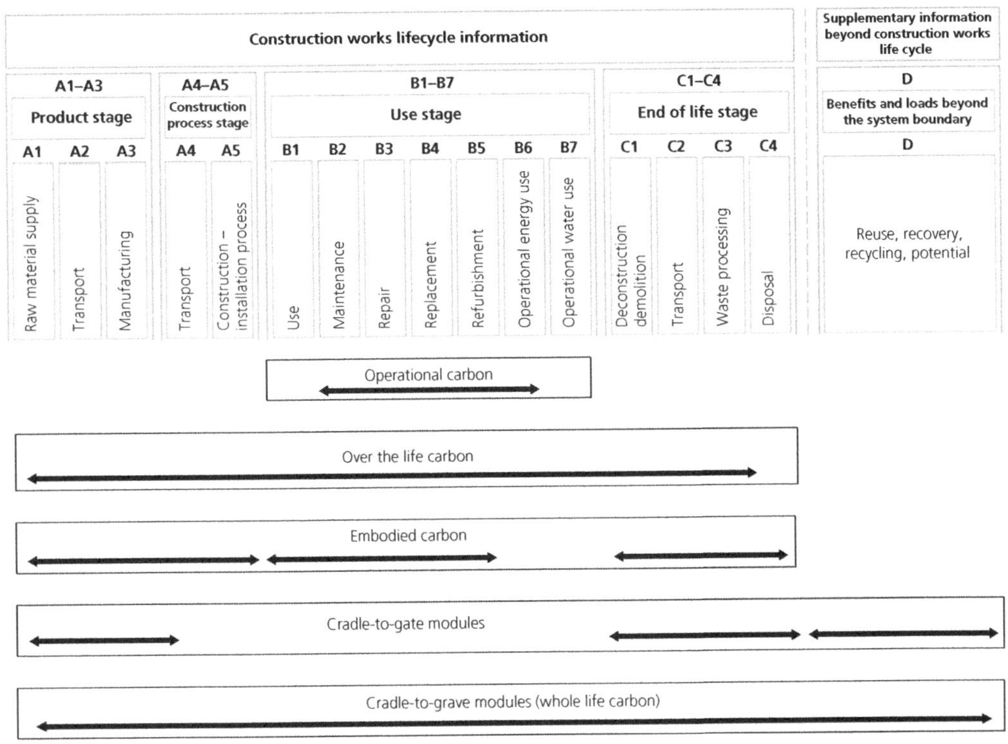

(physically integrated with other products and no longer identifiable at end of life and does not contain biogenic carbon), construction products and materials need to declare modules A1–A3, modules C1–C4, and module D. The mandatory stage that needs to be declared for cradle-to-gate using declared unit and cradle-to-grave using functional or declared unit is shown in Figure 5.18. Here functional unit relates to a specific function and scenario and the declared unit relates to a range of functionalities and scenarios. To ensure future emission and consumption at a minimum level, overall whole life EC assessment taking A–D into consideration is important. For example, using the ICE database, a total EC of 286 $kgCO_2e/m^3$ for lifecycle assessment (LCA) stages cradle-to-gate A1 to A3 with the exemption for modules C and D of structural concrete of design strength C25/30 was measured (Low Carbon Concrete Group, 2022). The breakdown of EC contributions among its constituents is illustrated in Figure 5.19, highlighting the distribution of emissions associated with each material component where cement is the major contributing factor (87%). However, the chosen cement type, which depends on specific project requirements for strength, durability, workability, and curing speed, predominantly determines the content of cement required to achieve a given concrete class, thereby largely influencing the EC. A similar EC calculation guide for structural elements for the built environment that aligns with BS EN 15978 (BSI, 2011b) and BS EN 15804 (BSI, 2022) has been published by Orr *et al.* (2022).

Figure 5.19 Distribution of EC in a typical structural concrete for LCA stages A1 to A3 cradle-to-gate. Adapted from Low Carbon Concrete Group (2022)

The EC of concrete or building fabric should be measured at the early stage of the design process when flexibility for making changes and adjustments within the different sections of the design team (such as materials, structure, building services, etc.) will create minimal disruptions in light of embodied carbon assessment.

Questions to practice

1. Considering the pozzolanic and hydraulic properties of supplementary cementitious materials, explain why the British Standard allows a higher replacement rate of Portland clinker with GGBS compared to fly ash.
2. Explain how Portland–fly ash cement CEM II/B-V can achieve greater reductions in the embodied carbon of structural concrete compared to CEM II/A-V. Additionally, appraise a key design principle that should be adopted to maximise the carbon reduction benefits of using CEM II/B-V cement. Here 'V' refers to fly ash.
3. Based on the physical and chemical properties of silica fume, critically evaluate its performance when used as a SCM or addition in concrete.

Monower Sadique
ISBN 978-1-83608-477-8
https://doi.org/10.1108/978-1-83608-474-720251006

Chapter 6
Alternative cement

After completing this chapter, you should be able to

- understand the concept of alternative binder systems and their potential to replace traditional Portland cement
- explore the principles of activation of alternative binders and how they influence their performance
- analyse how physical, chemical, and physicochemical activation methods can enhance the properties of alternative cementitious binders
- understand the fundamentals of alkali-activated cement (geopolymer binders) and their role as sustainable alternatives to traditional cement
- explore the functioning of calcium aluminate cement in systems that do not rely on Portland cement
- understand the properties and applications of magnesium oxide cement as an alternative binder
- examine the concept and performance characteristics of supersulfated cement in sustainable construction practices.

Introduction

Researchers and manufacturers worldwide are developing alternative cementitious systems to reduce reliance on traditional Portland clinker. The upward trend of utilising SCM and the subsequent consumption of cement replacement materials demands further research in order to explore the novel source. Blending different binders to harmonise their existing physicochemical properties through utilising their potentiality in a controllable system has been found to play a pivotal role in developing a novel cementitious material. Developments in the utilisation of fly ash, GGBS, and SF as SCM has been researched widely using various activation techniques that bring out different degrees of modification in terms of rheological and morphological properties of the subjected materials, and the successful application is a function of the materials' basic properties. Based on their differing activation techniques, various alternative low carbon cementitious systems are in use. This includes alkali-activated, calcium–alumina, magnesium oxide, and sulfate-based cementitious systems, which differ from conventional calcium-silicate-based Portland clinker. The primary motivation behind these alternatives is to minimise, if not eliminate, the calcination and process emissions of CO_2.

Alternative binder system (ABS)

Flex 350 (BSI, 2024) defines ABS as a 'substance formed by a combination of one or more constituent materials together with chemicals or activators which react to form a hardened monolithic material'. In addition to 31 products containing SCM specified in BS EN 197-1 & 5 (BSI, 2019a, 2021d), there are other innovative types of ABS in current use in the construction sector, of which AAC, calcium aluminate cement (CAC), magnesium oxide (MgO) cement, and supersulfated cement are worth mentioning. The following sections depict these types of ABS concrete.

Activation of alternative materials

Activation is essential for the effective utilisation of alternative materials or SCM. Three widely recognised activation techniques to enhance SCM reactivity include chemical activation, mechanical activation, and physicochemical activation. Various chemical substances such as alkaline reagents and alkali salts are widely used to excite the potential reactivity of SCM. Mechanical activation is intended to improve the degree of fineness through grinding. Physicochemical activation is achieved through a calcination and hydrothermal process (Wang and Wang, 2004). The reasons for lower activity of a potential SCM in general arise from two factors (Fan *et al.*, 1999).

- The dense and chemically stable glassy surface layer of glass beads protects the inside constituents, which are porous, spongy, and amorphous, and, therefore, have advanced activity.
- The silica–alumina glassy chain of ashes is rich in Si, Al, and low Ca is firm and the chain must be disintegrated if activity is to take place. With addition of $Ca(OH)_2$ and higher basicity the stable outer layer is corroded and the active core is exposed for further hydration reaction, producing cementitious material.

Globally, different potential calcareous and/or pozzolanic materials have been subjected to various activation processes for their utilisation as SCM. These are reviewed below, focusing on the key developments.

Chemical activation

There is a large volume of published studies describing the role and performance of alkali in cementitious systems. The chemical or alkali activation is a particular procedure by which the powder is mixed with specific alkaline activators (alkaline solutions) and then the mixture is cured under a certain temperature to make solid materials. High pH accelerates the breakdown of the glassy phase; $pH > 13.3$ may be desirable. In general, the main reaction of any pozzolanic material is the breakdown of the glassy part by attack from OH^- ions on the SiO_2 and AlO_2–SiO_2 framework (Jones *et al.*, 2004). The glassy constituent of the fly ash is changed into well-compacted cement. In previous works, it was found that the main reaction product formed in alkali activation is an amorphous alumina–silicate gel. This product, also considered a 'zeolite precursor', is an X-ray amorphous material (Fernandez-Jimenez and Palomo, 2005). In the previous research, various chemical activators such as lime, nano-SiO_2, Na_2OSiO_2, NaOH, Na_2SO_4, BFS, $CaCO_3$, and flue gas desulfurisation (FGD) gypsum have been used to activate fly ash materials. Qian *et al.* (2001) reported a more beneficial effect of activation of fly ash with Na_2SO_4 rather than grinding. Moreover, the combination of grinding and the addition of Na_2SO_4 gave higher strength than either the grinding or the addition of Na_2SO_4 alone. The incorporation of Na_2SO_4 increases the concentration of SO_4^{2-} and pH which results in the acceleration of the pozzolanic reaction as well as the formation of ettringite.

Mechanical activation

Induction of solid state reaction during grinding was reported by Lin and Somasundaran (1972) where the most prominent reaction was identified in the case of carbonates. Various types of milling devices such as attrition mills, ball mills, agitator mills, jet mills, vibro mills, and so on have been utilised for mechanical activation through grinding. Various factors such as initial particle size, softness, and carbon content affect the performance of grinding. Loss of effectiveness in grinding fly ash after 20 min in a laboratory ball mill was reported by Payá *et al.* (1995). The agglomeration of particles, where particles adhere to each other under the action of strong chemical bond forces, and particles with metastable high state for prolonged grinding consequently decrease the grinding efficiency.

A previous study by Bouzoubaâ *et al.* (1997) showed that the morphology of fly ash altered for grinding in such a way that it increased the mortar cube compressive strength of the cement–fly ash system. Sekulic *et al.* (2004) found that a compressive strength 3.3% higher than the control was achieved through grinding clinker with 20% additions. Furthermore, significant technical benefits were achieved by separating ultra-fine PFA particles using air-cyclonic processing (Jones *et al.*, 2006). However, this type of processing was not feasible considering the disposal of coarse fly ash fractions. Micronising fly ash particles by grinding was reported to be more effective than fractionating in terms of generating strength (Eymael and Cornelissen, 1996). The advantages of grinding listed by Lu and Wei (1992) were low cost, high output, and a simple process. A study in the UK found that the pH of paper sludge ash (PSA) slurry samples increases by milling the slurry with a drum and pebble mill, indicating that more free lime is present for hydration in slurry (Mozaffari *et al.*, 2009). Due to improved reactivity, the resulting cement and/or concrete exhibits improved strength and setting properties. Increased hydration of C_3S in clinker, early formation and consumption of CH, and more a compact structure with increasing levels of slag in the cement–slag system have been observed for grinding clinker and slag separately using an attrition mill, where milling was carried out with different durations ranging from 3 minutes to 60 minutes. But a sharp decrease in slag median size (d_{50}) was observed only after 5 minutes, and beyond 15 minutes the rate of decrease in particle size was very low and the lowest value of zeta potential was also observed at this stage (Kumar *et al.*, 2008). A similar observation was also reported during grinding 20% fly ash and 75% clinker in addition to 5% gypsum, and compressive strength equal to control after 28 days was achieved using a vibro mill (Sekulic *et al.*, 1999). Grinding sugar cane bagasse ash (SCBA) was also reported to provide more nucleation sites at ITZ through consumption of C-H and formation of secondary C-S-H and calcium aluminate silicate hydrates (C-(A)-S-H), resulting in enhanced compressive strength in addition to filler effect by refining the pore structure over time (Torres de Sande *et al.*, 2022).

Physicochemical activation

Formation of amorphous $CaO–Al_2O_3–SiO_2$ by decomposition of original inorganic phases for heating paper sludge up to 800°C was reported by Hojamberdiev *et al.* (2008). A similarly accelerated pozzolanic reaction for the hydrothermal process was reported by Wang *et al.* (2007) while curing the paste of 50% CKD and 50% fly ash at 38°C.

A study by Kadhim *et al.* (2020) reported a significant reduction and eventual disappearance of crystalline peaks, accompanied by a transformation to amorphous phases, in natural pozzolan and metakaolin when calcined at 950°C. Additionally, the calcination process led to the formation of new compounds composed of the same elements in lime kiln dust (LKD), further enhancing its reactivity. Besides calcination and hydrothermal operation, physicochemical activation can be achieved by blending different types of minerals with varying chemical and physical properties. Chemical (pozzolanic) and physical mechanisms were reported to be initiated for synergic action achieved through ternary blending of two types of fly ash (high and low calcium) and cement (Antiohos *et al.*, 2007). Moreover, physicochemical activation can be triggered by grinding where, in addition to the creation of finer particles, noticeable modification of differentiated thermogravimetry (DTG) profiles and formation of surface oxides can be ensured, which in turn is indicative of enhanced reactivity (Blanco *et al.*, 2005; Senneca *et al.*, 2011). Higher strength development of kaolin-based geopolymer and blast furnace cement for elevated curing temperature demonstrates the physicochemical activation through the hydrothermal process (Abd-El.Aziz *et al.*, 2012; Heah *et al.*, 2011). Gypsum-aided grinding of high lime fly ashes using mortar and pestle was found to create a binder with gap graded in addition to reduced particle sizes that positively influence the water demand, particle packing, and hydration behaviour (Sadique and Al-Nageim, 2012).

Alkali-activated cement (AAC)

AAC has attracted increasing attention due to its potential for reduced carbon emissions and enhanced durability (Hamed *et al.*, 2024; Provis and Van Deventer, 2013; Singh *et al.*, 2015a). Currently, there are no specific British Standards for alkali-activated concrete or geopolymer concrete (GC). However, recognising the growing interest in its application for developing low-carbon concrete, BSI has published PAS 8820 (BSI, 2016b), which provides guidance on the use of alkali-activated cementitious materials (AACM) in concrete applications. PAS 8820 defines AAC as a 'substance consisting of an alkali activator and an AACM powder, or blend of such powders, with or without the inclusion of subsidiary constituents, with or without the incorporation of Portland cement, which under aqueous conditions, reacts to produce a hardened monolithic material' (BSI, 2016b). AAC is synthesised through alkali activation/geopolymerisation, and the resulting product has excellent features, such as high early strength and high resistance to elevated temperature and aggressive conditions.

Alkali activation involves a reaction between solid materials that are rich in highly reactive amorphous or semi-amorphous silica and alumina with an alkaline activator to formulate a strong tri-dimensional alumina–silicate polymeric network (Davidovits, 1993). Most used alkali activators are silicates of sodium, potassium and sodium hydroxide, potassium hydroxide, sodium carbonate, and sodium sulfate. The mechanism of alkali activation depends on precursor materials and the type and molarity of the activators used. Once the Al–Si precursor powder is mixed with the alkaline agent (either aqueous or dry state), a paste forms through geo-polymerisation and rapidly converts into a hard, activated cementitious product with accelerated setting and a rapid rate of strength gain. The physical and mechanical requirements for AACM are set out in PAS 8820 (BSI, 2016b), while the conformity testing for the AACM concrete shall be carried out in accordance with BS EN 206 (BSI, 2021c). The source precursor materials for the AACM are specified as GGBS, fly ash, pozzolans, limestone, burnt shale, and silica fume. Figure 6.1 shows the typical appearance of demoulded conventional and geopolymer concrete 24 hours after casting.

Like conventional concrete, the curing process significantly shapes the activation of aluminosilicates and the final characteristics of geopolymer concrete (Akduman *et al.*, 2024; Gaurav *et al.*, 2024; Hamed *et al.*, 2024; Kanagaraj *et al.*, 2024; Shoaei *et al.*, 2024; Swathi and Vidjeapriya, 2023). The effect of the curing regime on conventional and fly ash–GGBS-based geopolymer concrete made with typical mix design parameters tabulated in Table 6.1 is illustrated in Figure 6.2. Ambient air-cured AAC exhibited the highest compressive strength at all ages when compared to water-cured control concrete, as well as AAC samples cured in ambient or hot water, while a slight decrease in strength was observed at 14 days which was recovered beyond this period. Irum and Shabbir (2024) reported 15–20% higher compressive strength in AAC samples cured in an oven at a temperature of 80°C for the first 24 h followed by ambient air-curing compared to the mixes cured under ambient air-curing conditions for enhanced dissolution and polymerisation processes.

Figure 6.1 (a) Normal concrete; (b) geopolymer concrete immediately after demoulding (author's own)

Table 6.1 Constituents used for traditional control concrete and AAC to investigate the influence of curing on strength

Concrete constituent	Fly ash	GGBS	CEM II/A-L	Na_2SiO_3	4M NaOH	Binder: fine aggregate	Fine aggregate: coarse aggregate	W/C ratio
Mix design	Binder composition: %			Activator/binder ratio = 0.35				
Geopolymer concrete (GC)	40	60	0	1	2	1:1	3:7	N/A
Control concrete	0	0	100	N/A	N/A	1:1	3:7	0.35

The author conducted an evaluation of geopolymerisation at the microstructural level under air and water curing conditions for an alkali-activated binder where GGBS and fly ash content were partially substituted with PSA and calcium carbide residue (CCR). CCR, a byproduct of acetylene gas production, has a pH above 12 and primarily consists of $Ca(OH)_2$. The optimal blend was identified as 30% GGBS, 25% fly ash, 30% PSA, and 15% CCR, with a Na_2SiO_3/NaOH ratio of 2:1. Mortar samples cured under both ambient air and water conditions exhibited similar strengths at early and standard durations. Microstructures of single-component AAC paste after 3 and 28 days of ambient water and air curing are shown in Figures 6.3 and 6.4. After 3 days, the air-cured samples showed more unreacted fly ash particles, while the water-cured samples exhibited greater geopolymer slurry formation. By 28 days, both curing conditions displayed a dense microstructure with enriched hydration products, comparable to those found in conventional fly ash–GGBS-based geopolymers reported by Kang *et al.* (2024). The appearance of cracks after 28 days of the water-cured sample was likely due to gel drying shrinkage.

In evaluating the diverse curing regimes (such as room temperature water curing, partially hot water curing, or air curing) for geopolymer concrete, each method offers unique advantages that can be strategically aligned with specific project requirements. Tailoring the choice of curing method to the desired outcome, whether it be cost-effectiveness, ease of manufacturing, or achieving optimal compressive strength, is essential for the successful application of geopolymer concrete in various

Figure 6.2 Strength development profile of AAC under different curing regimes (author's own)

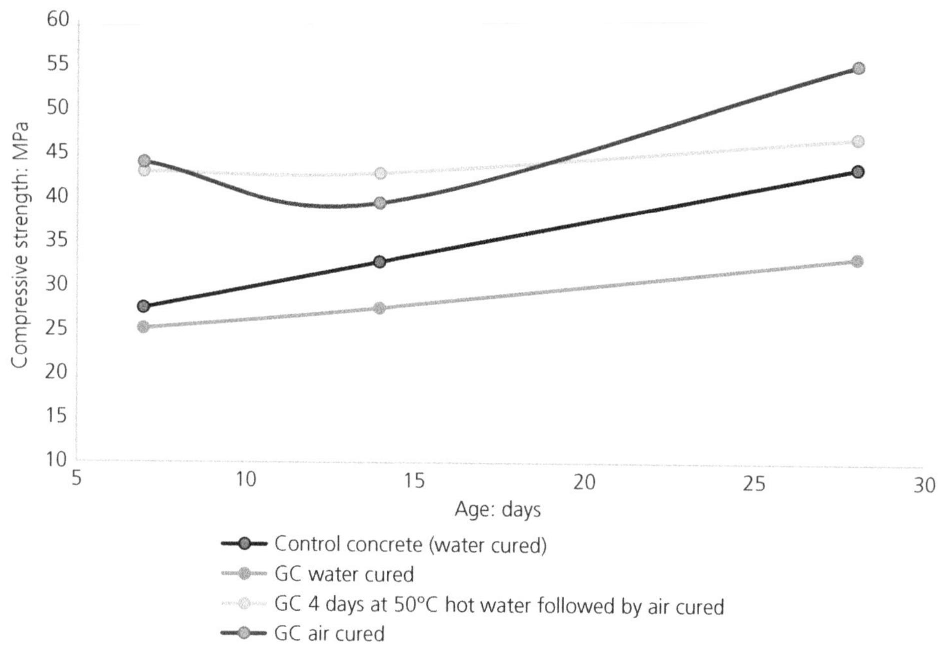

contexts. The selection of an optimal curing regime for geopolymer concrete should be guided by a comprehensive understanding of project-specific requirements. Balancing factors such as the nature of construction (cast in situ or precast), cost, time constraints, and sustainability goals will enable designers to make informed decisions, ensuring the successful application of geopolymer concrete in diverse construction scenarios.

Due to availability, proven alkali activation, and high content of Al and Si, GGBS and fly ash remain the most promising raw materials that have been used in AAC or in low-carbon cement alternatives so far. Wagners executed a record-breaking continuous pour of 737 m^3 of geopolymer concrete to fill the base of a 55 m-deep tunnel drive shaft at National Grid's Hurst Substation in South London, as shown in Figure 6.5. This fly ash- and GGBS-based system achieved a 64% reduction in carbon emissions compared to conventional concrete that would have traditionally been used (National Grid, 2023). However, globally as well as in the UK, the production of fly ash and GGBS will decrease significantly due to new regulations to control uses of coal in power generation and processing iron ore in blast furnaces to achieve the carbon target by 2050. This forms the main obstacle to the future application of AAC and composite cement technology in construction, especially in the UK (Duxson and Provis, 2008; Heath *et al.*, 2013; Heidrich *et al.*, 2015; Van Deventer *et al.*, 2012). The successful development and adoption of alternative types of cement or AACMs, such as those based on calcined clays or volcanic ash, have been recommended in the ICE *Low Carbon Concrete Routemap* to reduce dependency on GGBS and fly ash (Low Carbon Concrete Group, 2022). One promising alternative is metakaolin (MK), an artificial pozzolanic material produced by calcining kaolin [$Si_2O_5Al_2(OH)_4$] clay at high temperatures, which leads to the dehydroxylation of the OH groups. This process transforms MK into a widely adopted alumino-silicate source for low-carbon binders. Metakaolin, with its high SiO_2 and Al_2O_3 content, offers a stable chemical

Figure 6.3 SEM image of water-cured AAC materials containing GGBS, fly ash, PSA, and CCR: (a) after 3 days and (b) after 28 days (author's own)

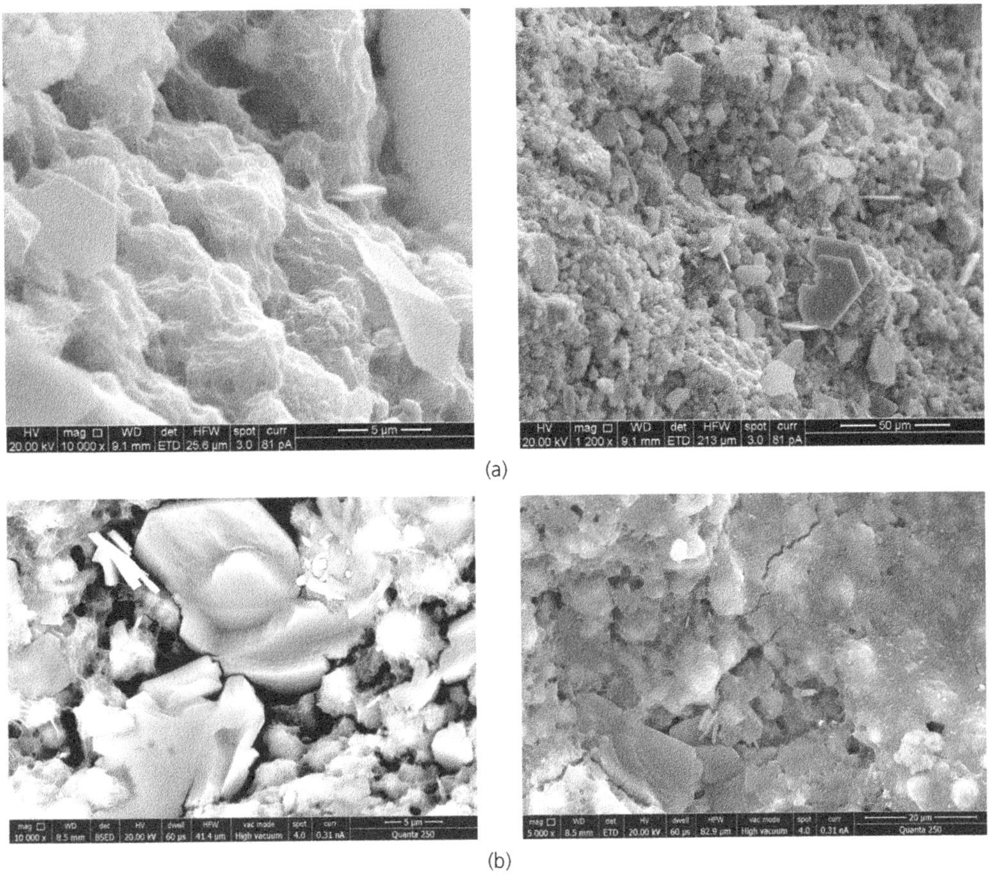

composition and benefits from a straightforward manufacturing process, ensuring consistency and predictability. Global estimated kaolin reserves are around 32 billion tonnes, mainly in the USA, UK, Ukraine, and Brazil (Wu *et al.*, 2016; Zheng *et al.*, 2022), making MK one of the most promising feedstocks for future geopolymer and low-carbon concrete development.

Calcium aluminate cement (CAC)

CAC, or high alumina cement, is a non-Portland binder primarily composed of calcium aluminate and low lime content, unlike Portland cement, which is calcium silicate-based. The raw materials used in the production of CAC include limestone, silica, and bauxite, the latter providing the essential alumina content. The composition, specifications, and conformity criteria of CAC are controlled by BS EN 14647 (BSI, 2005a), and its standard designation is identified as calcium aluminate cement EN 14647 CAC. Typically, the elemental composition of CAC consists of approximately 40% Al_2O_3, 20% Fe_2O_3, and 5% SiO_2, along with other minor compounds. The principal phase in CAC is monocalcium aluminate ($CaO \cdot Al_2O_3$), with other phases including C_2S, C_2AS, and C_6A_4FS. The manufacturing process of CAC is similar to Portland clinker, where the raw materials are sintered in a kiln to around 1600°C, then cooled and ground into a fine inorganic

Figure 6.4 SEM image of air-cured AAC materials containing GGBS, fly ash, PSA, and CCR: (a) after 3 days and (b) after 28 days (author's own)

(a)

(b)

hydraulic powder. Although the setting time of CAC is longer than Portland cement, the strength generation after both early and standard duration is rapid and higher than Portland cement. Due to its rapid hardening feature, CAC is widely used in manufacturing pre-stressed precast concrete structural elements, temporary tunnel linings, flooring, and abrasion-resistant surfaces, as well as rapid repair works, in addition to its inherent application in resisting sulphate attack. Relating to durability, CAC also offers resistance to ASR and high temperature, while its excellent acid resistance makes it well-suited for applications in sewer networks. It is important to note that the high heat of hydration generated due to CAC's rapid early strength development limits its use in mass concrete applications.

Magnesium oxide (MgO) cement

The dead-burned form of MgO present in Portland clinker, originating from the raw meal, hydrates slowly in hardened concrete, forming magnesium hydroxide, which causes expansion and unsoundness. Due to this detrimental effect, BS EN 197-1 (BSI, 2019a) limits the MgO content in Portland clinker to a maximum of 5% by weight. However, the beneficial application of engineered reactive MgO has attracted interest in developing an MgO-based alternative binder system. The major

Figure 6.5 Construction of 55 m-deep tunnel drive shaft at National Grid's Hurst Substation in South London using geopolymer concrete to fill the base. Courtesy of National Grid (2023)

elemental constituent of this type of cement is MgO or magnesia which can be used on its own or in combination with cement or SCM offering cementitious properties. The most widely used raw material for manufacturing MgO cement is the calcination of synthetic brucite or the naturally occurring mineral magnesite or $MgCO_3$ at 1000–2000°C in the kiln. The sintering process also releases CO_2 gas, as with Portland clinker. Lower temperature calcination around 650–1000°C produces the most reactive MgO, whereas fused MgO which is less reactive is generated at 2000°C. The calcined magnesite creates decomposed burned reactive MgO possessing higher porosity, lower crystallinity, and high surface area compared to other grades of MgO. The reactivity of MgO is largely dependent on its specific surface area, particle size, and surface defects which are influenced by the production conditions (raw meal, calcination temperature, and duration).

Unlike hydrated MgO, the precipitation of stable $MgCO_3$ is an established process achieved through the natural carbonation of MgO, which involves sequestering CO_2 gas in magnesium-rich solutions, such as waste brines from desalination or seawater. The CO_2 concentration and temperature influence the formation of $MgCO_3$ and many researchers (Botha and Strydom, 2001; Hänchen *et al.*, 2008) have tried to investigate the optimum condition for enhanced carbonation. Given the global scarcity of natural magnesite deposits and the carbon-intensive calcination process it requires like Portland clinker, the carbonation route of MgO is emerging as a more sustainable alternative for MgO cement production.

The rate of hydration and the binding ability of the resulting magnesium carbonates are influenced by the reactivity of the MgO used in the process. However, reactive MgO has a higher water demand in the region of 0.45–0.70 (Al-Tabbaa, 2013) due to its higher surface area and lower particle size compared to Portland cement. During hydration, reactive MgO precipitates as brucite ($Mg(OH)_2$), which is subsequently carbonated by reacting with dissolved CO_2, converting into

Figure 6.6 Strength development of MgO concrete containing hydration agents under (a) ambient and (b) accelerated carbonation curing (Dung and Unluer, 2017). Courtesy of Elsevier

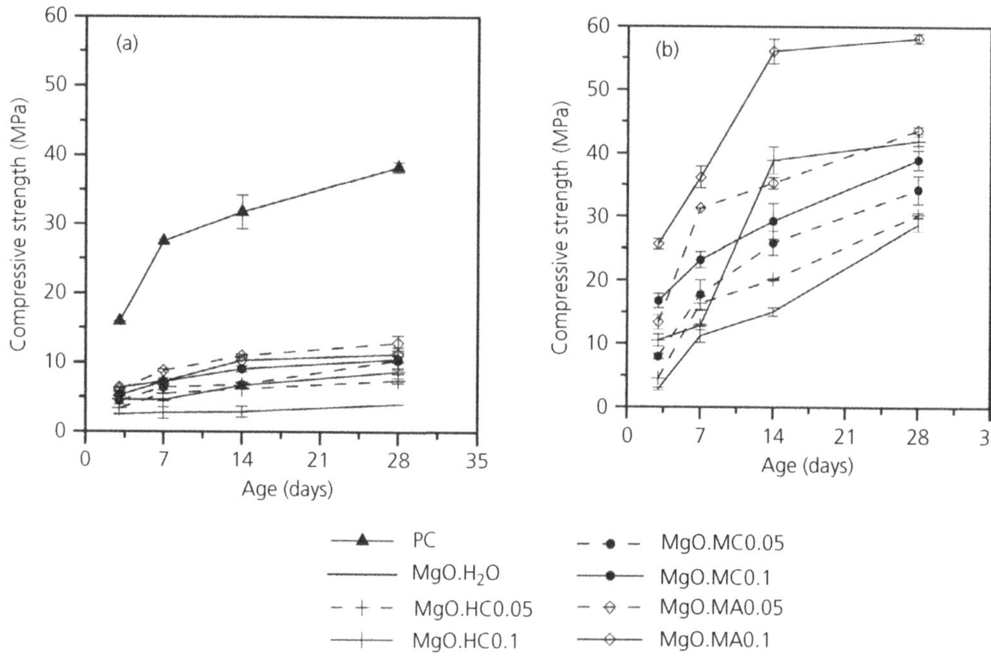

stable hydrated magnesium carbonates with limited cementitious properties. Therefore, the degree of hydration as well as carbonation play a crucial role in determining the binding capacity of the hydrated MgO cement. A combined application of the different concentrations (0.005 and 0.1 M) of hydration agents (magnesium acetate (MA), magnesium chloride (MC), and hydrochloric acid (HC)) and accelerated carbonation (30°C, 80 ± 5% RH, 10% CO_2) curing were found to increase the 28-day compressive strength by 14 times (56 MPa) relative to the control MgO samples cured under the ambient condition (4 MPa), as shown in Figure 6.6 (Dung and Unluer, 2017).

For carbonated MgO, the rate of strength gain is primarily driven by the carbonation process rather than hydration, in contrast to hydrated MgO. Increased densification and greater micro-hardness due to carbonation curing for 56 days compared to non-carbonated samples in the MgO–Portland cement system for the formation of both calcium and magnesium carbonate were reported by Mo and Panesar (2012).

The enhanced resistance of hydrated MgO cement against sulphate attack has led to its specialised application in stabilising soil contaminated with high sulphate and heavy metals. The potential application in the case of superior resistance to acid and sulphate attack by MgO porous block compared to Portland cement has been suggested by Liska and Al-Tabbaa (2012). The application of pure MgO as an expansive additive due to the formation of brucite with increased volume to control thermal cracking of mass concrete was first recorded in China in 1970, and wet curing and increased temperature were found to enhance the expansive ability (Liska *et al.*, 2019). The utilisation of other magnesium-rich industrial wastes, such as high-magnesium nickel slag, has also been explored in developing geopolymer concrete (Bouaissi *et al.*, 2019). This process was found to form morphologically less porous M-S-H or C-M-S-H gels, which significantly enhanced the mechanical properties and improved the microstructural characteristics.

Supersulfated cement (SSC)

Unlike CAC and magnesium oxide (MgO) cement, SSC or sulphate-activated slag cement is a calcination-free binder where GGBS is activated using gypsum, sodium, or potassium sulphate. In the presence of water, the sulphate ions initiate the hydration of GGBS, resulting in the formation of key hydrates ettringite (AF_t) and calcium silicate hydrate (C-S-H). To enhance early strength development, a small amount of alkaline activator using clinker or lime is often added during SSC production. The rate of hydration and the associated heat evolution in SSC are significantly lower compared to Portland cement. The significant influence of the chemical composition of GGBS over the amount of glassy phase present in the slag was found when low- and high-reactive GGBS were compared (Gruskovnjak *et al.*, 2008). As shown in Figure 6.7, high-reactive slag, characterised by

Figure 6.7 Calculated volume percentage of the solids formed in (a) high-reactive SSC and (b) low-reactive SSC as a function of the amount of GGBS dissolved (Gruskovnjak *et al.*, 2008)

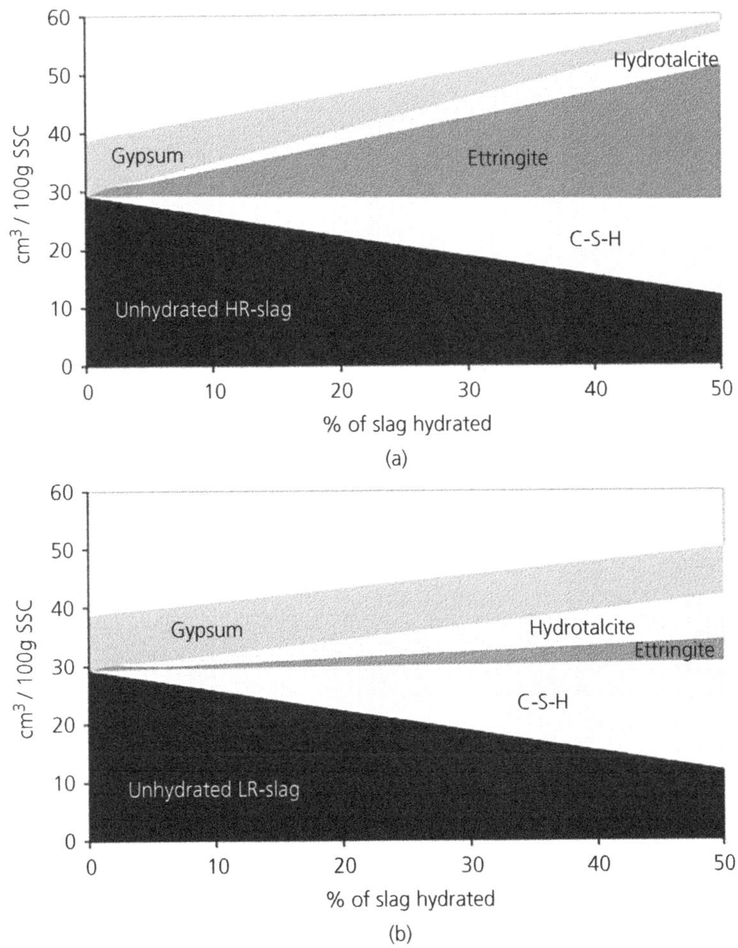

higher Al_2O_3 and lower MgO content, enhances the intrinsic dissolution rate. This results in the formation of more ettringite, leading to a greater volume of hydrates with reduced porosity.

In addition to its low heat of hydration, SSC offers better resistance to sulphate attack (due to the absence of C-H) and remarkable performance in minimising drying shrinkage, making it a highly durable and reliable material for various construction applications. As a higher amount of ettringite develops, a higher overall porosity in SSC compared to Portland and GGBS–Portland binder is reported (Liu *et al.*, 2014). However, the formation of more gel micropores (< 4.5 nm) and fewer medium capillary pores (50–100 nm) resulting in a lower volume of harmful pores in hydrated SSC is reported to improve the strength and impermeability (Mehta and Monteiro, 2014). The low alkalinity and low resistance against carbonation due to the ease of neutralisation with CO_2 restrict the use of SSC in practice (Wu *et al.*, 2021). Additionally, since SSC production does not involve the calcination of raw materials, its carbon footprint is considerably reduced, making it a more environmentally sustainable alternative to Portland cement.

Monower Sadique
ISBN 978-1-83608-477-8
https://doi.org/10.1108/978-1-83608-474-720251007

Chapter 7
Recent trends in concrete

After completing this chapter, you should be able to

- understand the concept of self-activated geopolymer and its mechanism of action
- explore the potential of calcined clay to reduce dependency on natural limestone in cement production
- examine the process of manufacturing 3D concrete, its advancements, and applications
- understand the prospects and challenges of carbon capture and storage (CCS) in the concrete sector
- discover how biomass ash can serve as a viable replacement for cement
- explore the recent developments in utilising agricultural and industrial wastes, including paper sludge ash, rice husk ash, and calcium carbide residues, as sustainable cement replacement materials
- analyse how natural minerals like olivine and engineered materials such as carbon nanomaterials contribute to sustainable concrete production
- understand the production and benefits of special concretes, such as high-strength concrete, high-performance concrete, and self-compacting concrete, in achieving durability and sustainability
- investigate the role of artificial intelligence in driving innovation and efficiency in the concrete industry.

Introduction

The concrete sector relies significantly on low-carbon technologies using GGBS and fly ash as feed materials. However, the future supply of these essential materials is threatened by the phasing out of coal-burning power plants and blast furnaces in steel manufacturing, both in the UK and globally, to meet carbon reduction targets. Additionally, the health and safety risks posed by corrosive chemical activators in traditional geopolymer production in an otherwise promising clinker-free binder system limit its widespread in situ application. Consequently, alternative materials like calcined clay, biomass fly ash, alkaline industrial wastes, and so on, as well as advanced materials processing and manufacturing technologies like 3D printing, carbon sequestration, carbon nanomaterials, high-strength concrete, self-compacting concrete, artificial intelligence (AI), and so on, are under development as sustainable solutions.

Self-activated geopolymer

So far geopolymer research has focused on traditional two-part concrete where fly ash and GGBS-based alumina–silicate precursors are activated with concentrated aqueous solutions of alkali metal hydroxides and silicates. On 1 October 2024, the UK closed down its last 2000 megawatt (MW) coal plant in Nottinghamshire and became the first G7 country to end coal-fired power production, ending the UK's 142-year reliance on coal (BBC, 2024). Similarly, on 30 September 2024, Tata Steel, one of the UK's largest steel manufacturers, shut down its last blast furnace as part of its transition toward more sustainable, eco-friendly steel production methods (Croft, 2024). Therefore, regulations to decrease greenhouse gases by more than 80% by 2050 in many countries, especially in Europe and elsewhere, will significantly impact the production of fly ash and slag in the next few years. Moreover, conventional GC activated with an alkaline chemical solution is categorised as extremely corrosive, sticky (Lei *et al.*, 2024), and difficult/expensive to handle (health and safety concerns); activator solutions provide the highest single contributor to embodied CO_2 in GC (Luukkonen *et al.*, 2018; Provis *et al.*, 2015; Van Deventer *et al.*, 2012). Establishing the emissions associated with the production of alkaline activators as well as reducing the dependency on fly ash and GGBS was also emphasised in the ICE *Low Carbon Concrete Routemap* (Low Carbon Concrete Group, 2022). The possibility of using any form of GC in large-scale applications has not been vindicated in the construction sector widely. Non-availability of real-site case studies or performance data of GC in large-scale structures means designers are reluctant to specify it. Heath *et al.* (2013) reported that acceptance is expected to be improved if GC is formulated in a similar manner to conventional concretes through mix design (no chemical activator) and through reinforcement.

To address the issues relating to alternative Al–Si sources and avoidance of alkaline chemical activators, several efforts have been made under the name of one-part geopolymer, self-activated geopolymer, or one-part alkali alumina–silicate binder where only dry powder is mixed with water like conventional cement. The use of GGBS and biomass ash derived from the co-combustion of rice husk and corn straw, as a substitute for fly ash, activated with sodium carbonate and sodium silicate, resulted in a compressive strength of 46.09 MPa at 28 days, along with a reduced setting time (Lei *et al.*, 2024).

Kadhim *et al.* (2021) reported the innovative use of thermally activated LKD waste as an alkaline activator to develop a one-part self-activated geopolymer binder, utilising MK and natural pozzolan as alternatives to fly ash and GGBS. This approach demonstrated enhanced reactivity and activation of aluminosilicate minerals through LKD, which breaks down crystalline phases and facilitates the formation of new wollastonite minerals after calcination. This advancement represents a significant step forward in addressing supply chain limitations and mitigating the health hazards associated with traditional geopolymer concrete technology.

Calcined clay

Natural calcined pozzolana (Q) is one of the two sub-categories under pozzolanic materials specified as SCM in BS EN 197-1 BSI (2019a) as materials of volcanic origin, clays, shales, or sedimentary rocks, activated by thermal treatment or calcination. Successful disruption of the crystal and inert structure of clay by calcination and associated enhancement of reactivity and pozzolanicity have attracted considerable interest where scarcity of widely used SCMs such as fly ash and GGBS is imminent. Adopting new technology and identification of clays with mineralogy suitable for calcining to use as SCM has also been suggested to reduce the carbon intensity in the ICE *Low Carbon Concrete Routemap* (Low Carbon Concrete Group, 2022). The carbon intensity of calcined clay is high as it requires high temperatures for calcination; however, unlike Portland clinker, it does not have any process emission. Calcination of kaolinite clay ($Al_2Si_2O_5(OH)_4$), also known as China clay, in the range of 550–800°C for 2 h produces highly reactive amorphous metakaolin

($Al_2O_3 \times 2SiO_2$) rich in aluminium and silicon oxide through dihydroxylation. The purity (crystal structure, size, etc.) of the kaolinite clay, calcination temperature, duration, and cooling rate influence the reactivity of metakaolin. MK with less reactivity for calcining below 700°C (Ambroise *et al.*, 1985) and further crystal formation followed by a decline in reactivity for calcining above 850°C (Marwan *et al.*, 1992) have been reported. Portland clinker replacement in the range of 6–35% by calcined pozzolan for Portland–pozzolana cement is allowable in BS EN 197-1 (BSI, 2019a). In the Portland cement (PC)–MK system, MK follows a conventional pozzolanic reaction and forms C-(A)-S-H; hydrated aluminium silicate fills the gap in addition to its filler effect and contributes to strength and durability properties. Although increased cement replacement levels with MK have been reported to reduce concrete workability and extend setting times, the use of sulfonated vinyl copolymer superplasticisers has been shown to counteract these effects by enhancing compressive strength (Brooks and Megat Johari, 2001). The enhanced initial setting time for MK addition is beneficial for the mechanical and transport properties of the concrete. The reactivity of MK with high water demand has attracted its application in alkali-activated materials such as fly ash–MK (Qiao *et al.*, 2024) and GGBS–MK (Azar *et al.*, 2023, Davidovits, 2008) based geopolymer rather than in the PC–MK system.

Given the limited global availability and supply chain challenges of high-purity kaolinite clay, which is essential for producing metakaolin, there is growing interest in the calcination of alternative clays, such as London clay and oil shale. These materials are being explored for their potential to be valorised and used as SCMs, similar to metakaolin, offering a viable solution in regions where kaolinite deposits are scarce, unlike more abundant resources such as limestone. London clay, excavated from the Crossrail tunnelling project, was calcined at temperatures ranging from 600–1000°C for its potential use as an SCM with CEM I (Zhou *et al.*, 2017). Calcination at temperatures of at least 900°C for 2 h was found to be necessary to achieve complete dihydroxylation, optimising the reactivity of the clay for use in cementitious applications in this study. Replacing 30% CEM I with calcined London clay showed 28-day strength greater than the control (100% CEM I) at 0.3 w/b ratios, despite relatively low kaolinite content, as shown in Figure 7.1.

Additive layer manufacturing (ALM)

Three-dimensional concrete printing (3DCP) or ALM benefits (fluidity, shape stability, and lower concrete formwork pressure) over conventional concrete have attractions for the construction sector. ALM is developed with the progression of 3DCP which includes contour crafting, concrete printing, and D-shape. 3DCP deposits rheologically controlled viscous cementitious material through a printing nozzle to cast concrete elements. Two main components of 3DCP are the printer (printing system) and ink (cementitious materials) for the printer. The printer is pre-programmed to extrude ink through nozzles following a designed path and deposits ink to make concrete elements. Flowability, extrudability, buildability, open time, pumpability, printability, workability, constructability, and manufacturability are crucial aspects that need close monitoring during the 3DCP process (Chung *et al.*, 2021; Hasani and Dorafshan, 2024). However, the 3DCP developed to date is not concrete per se, but a mortar (max $\phi = 2$ mm sand) or paste with no aggregate ($\phi > 4$ mm). Any dimensional changes under load to the paste and porous structure of mortar limit its acceptance for use in structural load-bearing members. Cement paste and mortar are commonly used for small-volume applications for repair and brickworks, respectively, with the most important application being reinforced coarse-grained concrete structural members. No research has been published relating to conventional coarse-grained concrete in the ALM process due to reduced pumpability and extrudability. Therefore, a limited understanding of ALM performance in large-scale structural member(s) (bridge girder, long-span beam, retaining wall, slab, etc.) hinders application and acceptance. Adding reinforcement as part of the ALM concreting process is

Figure 7.1 Strength development of concrete specimens where 0–30% CEM I was replaced by calcined London clay (CLC) at different w/b ratios (Zhou *et al.*, 2017). Courtesy of Elsevier

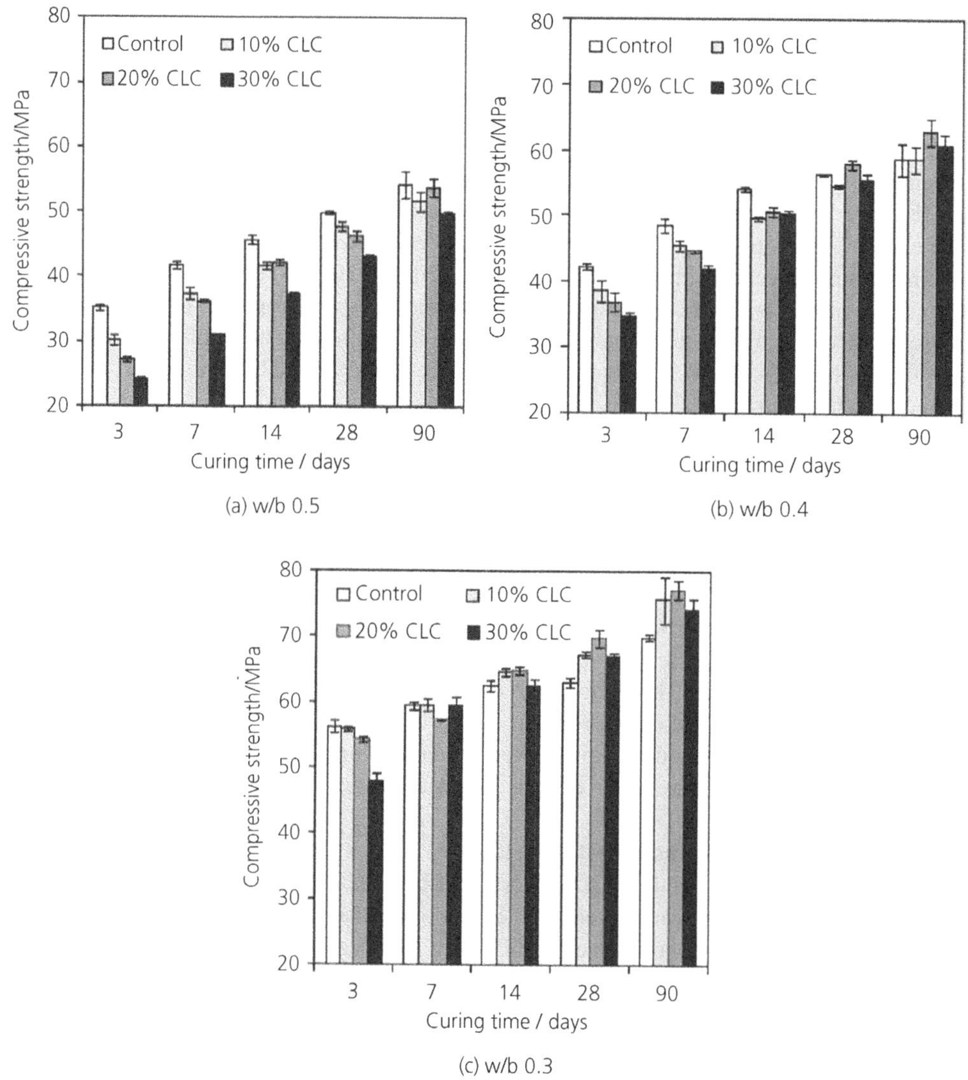

essential to make the product structural. Tensile reinforcement beyond the capacity of the concrete is essential for many applications; therefore, incorporation of reinforcement is necessary. Uncertainties relating to printing the encasement of reinforcement and the ability to manufacture reinforcement into novel or instantaneous in situ shapes have been reported by Buswell *et al.* (2018). Reinforcement solutions in contour crafting have been proposed by Khoshnevis (2004), including a screw-based system, but with no published performance characteristics of this system. The application of two-dimensional glass fibre-based, wide-mesh textile horizontally between the printed layers, covering the entire cross-section of the structural element has been reported by 3ders (www.3ders.org, 2014). However, the efficiency of the textile arrangement is unknown. A full-scale prototype column formwork to apply ALM for concrete building components using cement mortar with external dimensions of 0.60 m × 0.30 m × 2.90 m and 4 × 4 longitudinal steel reinforcement bars of 12 mm diameter was printed by Alabbasi *et al.* (2023), as shown in Figure 7.2.

To date, several different printer platforms based on gantry systems and robotic arms have been used for onsite and offsite 3D concrete printing – for example, in a gantry-based 3D concrete printer, an in-house customised LabVIEW-controlled 3D printer based on a three-axis flatbed configuration can be used with a 200 x 100 x 60 cm build volume, 0.5–1 mm Z-axis resolution and a speed of 200 mm/s on the XY-plane. The concrete workpieces (beam and slab) can be printed on a removable trolley under the XYZ-stage platform. Concrete is pumped from a material delivery system that can be extruded from the nozzles with different sizes and shapes. Proximity sensors, pressure sensors, and rheology sensors are integrated into the 3D concrete printer for position/motion validation, in-process monitoring, and quality control. The manipulation of a 3D printer mainly involves motion control and extrusion control of the printer head which are functions of layer thickness and product dimension. The printer and concrete workpieces are fed into the motion and extrusion control, enabling accurate control of header position and feed rate. It is essential to develop an advanced work planning strategy to achieve efficient and collision-free 3D printing. It has been demonstrated that good trajectory planning for the nozzle, including the planning of the nozzle path and its travelling speed, helps improve printing efficiency (Ganganath *et al.*, 2016). The in-process monitoring and dynamic control can address various constraints such as the different hardening times of the concrete and the additive material, the complex geometry of the target, and the effect of the external environment.

Carbon sequestration, capture, and storage

CCS is a recognised method of mitigating global carbon emissions while allowing continued use of fossil fuels. As discussed in the carbonation degradation of concrete in Chapter 4, atmospheric

Figure 7.2 3D-printed concrete column formwork and completed concrete with reinforcement (Alabbasi *et al.*, 2023)

CO_2 which penetrates in-service concrete combined with moisture creates carbonic acid and reacts with the calcium hydroxide (C-H), creating $CaCO_3$ (a strength-generating as well as pH-reducing agent), sequestering CO_2 that is permanently stored in the cement matrix. Carbonation curing of cement-based concrete is an alternative method to capture carbon than the geological storage approach. However, uptake through natural carbonation is very slow and dependent on the exposed surface area of the concrete and the concentration of CO_2 in the surrounding atmosphere. Possan *et al.* (2016)'s model predicts that while a building is in use, it will take at least 20–30 years to reach about 20 $kgCO_2/m^3$, equivalent to 5–10% of its upfront emission.

In recent years, injecting CO_2 or utilising accelerated carbonation curing with industrial CO_2 has gained attention for its advantages over traditional steam curing. Key factors influencing this process include the type of raw materials, water-to-cement ratio, preconditioning, CO_2 concentration, pressure, curing time, and subsequent curing methods (Shi *et al.*, 2017). The C_2S and C_3S phases of anhydrous cement as well as C-H and C-S-H (the hydration products) act as carbon sinks to capture carbon and create $CaCO_3$ during natural mineral carbonation. Further, the presence of AFt and AFm phases in hardened cement offers additional routes to carbonate formation. Curing reinforced concrete by accelerated carbonation at an early age where 11% CO_2 uptake based on cement content was not only found to improve concrete resistance to chloride-induced corrosion due to the formation of a densified carbonate-rich surface layer (Xian *et al.*, 2022) but also reduced porosity, drying shrinkage, and water absorption, while higher compressive strength with improved dimensional stability were reported by Shi *et al.* (2017).

In non-Portland binders, such as in MgO cement, the precipitation of stable $MgCO_3$ is an established process achieved through the natural carbonation of MgO, which involves sequestering CO_2 gas in magnesium-rich solutions. Likewise, mineral carbonation of olivine where the dissolution of magnesium silicates occurs resulting in $MgCO_3$ and amorphous silica (Wang *et al.*, 2019) has a higher potential application as SCM in concrete; thus carbon sequestration follows the reaction

$$Mg_2SiO_4 + 2CO_2 = 2MgCO_3 + SiO_2 \quad (7.1)$$

According to the Low Carbon Concrete Group (2022), 90% of the total carbon emitted during Portland cement production is due to burning fossil fuel and the decomposition of limestone powder which are unavoidable. Therefore, instead of using generated CO_2 expressly for this purpose, captured CO_2 during cement production or other industrial processes needs to be adopted to realise the true benefits of CCS technology. However, further advancements of this technology are required before there can be a large-scale roll-out in a feasible commercial environment (Low Carbon Concrete Group, 2022).

Biomass ash

Biomass is biological material derived from living or recently living organisms. Its significant reduction of net carbon emission compared with conventional fossil fuel makes biomass attractive for use as an energy source in the UK and worldwide. Based on the originating sources, biomass is categorised into the following five classes (Biomass Energy Centre, 2012).

- virgin wood: wood and other products such as bark, sawdust wood chips, and so on
- energy crops: high-yield crops grown for energy applications such as short rotation willow coppice, poplar, miscanthus grass, and so on
- agricultural residue: residue from agricultural harvesting such as straw, husk, corn stover, and poultry litter

- food waste: from food and drink manufacture, preparation, and post-consumer waste from hotels, restaurants, and individual houses
- industrial waste: from manufacturing and industrial processes such as paper pulp and waste, textile, sewage sludge, and so on.

Combustion of biomass results in two types of ash. These are (*a*) primarily larger particles that fall through the grate during combustion, known as biomass bottom ash, and (*b*) very fine particles that are carried in flue gases, known as biomass fly ash. Due to initiatives by environmental agencies to maximise the use of renewable energy sources, the use of biomass is increasing and the resulting wastes demand better management through further uses. Different types of BFA and BBA have been studied as potential pozzolanic material such as rice husk ash (RHA), SCBA, sugar cane straw ash (SCSA), sewage sludge ash (SSA), PSA, and so on.

A comparative study between RHA, SCBA, and SCSA conducted by Martirena Hernández *et al.* (1998) revealed that they have similar SiO_2, Al_2O_3, Fe_2O_3, and K_2O content but different CaO content. The main factors that drive the reactivity are the degree of crystallinity of the silica, unburned carbon content within the ash, and combustion process. In a lime–pozzolana system (30% lime–70% ash), SCSA was found to be similar to cement and RHA and more reactive than SCBA (Martirena Hernández *et al.*, 1998). Coarse particle sizes of SCBA are also detrimental for successive hydration, as compressive strength of mortar containing SCBA was found to be inversely proportional to its particle size. A study by Chusilp *et al.* (2009) concluded that ground SCBA showed greater compressive strength and lower permeability than the control concrete while 20% cement was replaced. Highest particle packing as well as advanced pozzolanic activity for mechanical activation by vibratory grinding of SCBA was also reported, concluding that SCBA may be classified as a pozzolanic material, but its activity depends significantly on its particle size and fineness (Cordeiro *et al.*, 2008). The ash of bagasse after combustion at 750–800°C was activated by mortar and pestle grinder and was reported to show a 62% increase in mortar compressive strength after 28 days, a decrease in porosity by 35%, and a reduction in water demand by 34.5% where fine aggregate was partially replaced by SCBA (Torres de Sande *et al.*, 2022). Moreover the refinement in the pore size distribution for combined blending effect of ground 10% RHA and 10% SCBA in a ternary system containing 60% cement was found to produce greater strength than control concrete (Cordeiro *et al.*, 2012). Due to extensive global research and the application of PSA and RHA as SCM, these two materials are discussed below.

Rice husk ash (RHA)

RHA is a reactive artificial pozzolanic material obtained by controlled calcination followed by grinding of rice husk after the rice grains have been separated. Due to the enormous global production of rice, the availability of rice husk and its sustainable management is a challenge. After calcination at moderate temperature, the resulting ash contains about 92–97% amorphous form of silica and exhibits high activity, which has resulted in its industrial application in silicon-based manufacturing and concrete sector. The calcination temperature and duration, as well as impurities attached to the rice husk, influence the amorphousness or reactivity of RHA, and its reactive siliceous content draws attention as an alternative to silica fume and nano-silica for high-performance concrete. In a fly ash–GGBS-based geopolymer concrete, GGBS and fly ash combinedly replaced by 15% RHA and reinforced with 2 vol% steel fibre showed around 159 MPa compressive strength with reduced porosity and enhanced interface bonding after 28 days in ultra-high-performance concrete (UHPC) (Pu *et al.*, 2024). Raw RHA exhibits a low activity index of 58% (< 75% required by

BS EN 450 (BSI, 2012a)) due to the presence of higher unburned carbon particles (Rukzon *et al.*, 2009); however, different activation methods such as sieving, grinding, and further calcination were found to improve its reactivity and enhanced its activity index. A decrease in water demand with increased grinding time and fineness of RHA, as well as enhanced filler effect and pozzolanic reaction, were reported by Rukzon *et al.* (2009). A 16% increase in the 28-day strength of mortar containing ground RHA compared to unground was reported for lesser amounts of unhydrated cement grains, air voids, $Ca(OH)_2$ content, and greater amounts of C-S-H gel (Kizhakkumodom Venkatanarayanan and Rangaraju, 2015). This improvement is attributed to the ground RHA's ability to reduce unhydrated cement grains, air voids, and $Ca(OH)_2$ content, while promoting a higher formation of calcium silicate hydrate (C-S-H) gel.

Paper sludge ash (PSA)

In 2023, $32\,659 \times 10^3$ tonnes of pulp was produced by 856 pulp and paper and board manufacturers in Europe, while around 180 million tonnes was produced globally (CEPI, 2024). Paper sludge consists mainly of wet, short, cellulose fibres lost during various stages of the papermaking process, the properties of which are dependent on the composition of the raw meal used in the process. Paper mill sludge is a major economic and environmental problem for the paper and board industry, as reported by Dunster (2007). The UK paper industry generates approximately 464 361 wet tonnes of waste sludge and 137 860 wet tonnes of rejects annually. Currently in the UK, 42% of paper sludge is repurposed as animal bedding, while the remaining 58% is used for land spreading. In contrast, the majority of paper rejects are sent to landfill sites (CPI, 2021). Current legislation, increased taxation, and commitment to climate change mitigation strategies have driven the need for innovation and research on non-conventional methods for the management or new uses of these solid wastes in more value-added products. In 2022, 62% of fuels used in the European pulp and paper industry came from biomass; a large amount of that is process residue (CEPI, 2024). Moreover fluidised bed boiler technology provides a means for successful thermal oxidation of high moisture wastes like paper sludge, producing process steam and/or electricity and reducing the amount of material to be landfilled by about 80–90% (Monte *et al.*, 2009). As per the CEPI sustainability report, the rate of landfill disposal was reduced from 32.3 kg/tonne to 15.2 kg/tonne of product from 2000 to 2010 (CEPI, 2011). As a result, coal and fossil fuel consumption has decreased, while the consumption of gas and biomass has increased in the production process of paper in Europe.

A study conducted by the Waste and Resources Action Programme in 2007 concluded that the most beneficial uses of the waste paper sludge are: (*a*) in cement; (*b*) as a brick additive; and (*c*) as a fuel substitute in a biomass boiler (WRAP, 2007). The compositional matrix of a pozzolanic or hydraulic material determines its activity during hydration. Being a tertiary waste, PSA poses different compositions based on its source of origin, combustion, and pollution control measures at the incineration plant. The change in composition for calcining paper sludge between 700°C and 1200°C is tabulated in Table 7.1. From Table 7.1, it can be stated that combusting paper sludge in the range of 700–1200°C did not greatly affect the composition, whereas the co-combustion materials and combustion system significantly changed the Ca and Si content. At the same time, the paper sludge source of origin also played an important role in this case. A consistent formation of hydration products was reported in a non-Portland binder using a ternary system of PSA, silica fume, and another high alkali fly ash where the stable and non-expansive nature of secondary ettringite was also identified after a long curing time (Sadique *et al.*, 2019).

Table 7.1 Chemical composition of PSA investigated by different researchers

	CaO	SiO_2	Al_2O_3	Fe_2O_3	K_2O	Na_2O	SO_3	Remarks	Source of origin
Banfill and Frias (2007)	31.4	30.2	18.0	0.7	0.32	0.21	0.27	Combusted at 700°C for 2 hours	Spain
Segui *et al.* (2012)	45.5	28.0	13.2	1.3	0.7	0.4	1.3	Paper sludge is co-combusted with wood waste at 850°C	France
Bai *et al.* (2003)	43.5	25.7	18.8	0.8	1.3	1.5	1.05	850°C and 1200°C and then cooled rapidly to 200°C to prevent the formation of dioxins	UK
Sadique *et al.* (2019)	57.0	28.0	3.7	0.2	–	–	0.3	Combustion between 700°C and 1200°C in power generation plants using a fluidised bed combustion system (Sadique *et al.*, 2012)	UK

Calcium carbide residue (CCR)

CCR, an industrial by-product of acetylene gas (C_2H_2) production, presents a significant waste management challenge for foundation industries reliant on calcium carbide (CaC_2). The global market for CaC_2 is projected to reach 33.25 million metric tonnes by 2024 (Mordor Intelligence, 2024), with approximately 1.15–1.45 tonnes of CCR generated per tonne of acetylene gas produced by the hydrolysis of CaC_2 (Ma *et al.*, 2016). This process results in an estimated annual global CCR waste generation of 56 million tonnes (Zhang *et al.*, 2016). For high alkalinity, currently landfill disposal remains the predominant method of CCR waste management worldwide, including in the UK and China. However, this approach contributes to significant environmental issues. The high alkalinity (pH in the range of 12.5) of CCR, primarily consisting of 80–95% $Ca(OH)_2$ with trace amounts of Si, Al, and Fe, presents a valuable opportunity for its utilisation as a SCM, offering a sustainable alternative to conventional waste management practices due to its potential reaction with high SiO_2- and Al_2O_3-containing materials producing cementitious hydrates.

After acetylene gas production, the resultant CCR slurry is compressed into cakes for disposal, as depicted in Figure 7.3(a). The dried and ground form of CCR is shown in Figure 7.3(b). The SEM image of the dry CCR powder in Figure 7.4 reveals a particle size comparable to that of cement, with particles exhibiting an almost spherical shape. Furthermore, the XRD analysis presented in Figure 7.5 confirms the predominant presence of portlandite as the primary compound, with minor traces of calcite.

While CCR has been successfully utilised at a low application rate as a filler in asphalt materials (Dulaimi *et al.*, 2020, 2021), its potential application in cementitious systems appears more promising. This is primarily due to its high alkalinity, substantial calcium content, and fine particle size, which collectively enhance its suitability as a supplementary cementitious material. Improvements in compressive strength and modulus of elasticity by replacing cement with CCR and SCBA was

Figure 7.3 Image of (a) as received dry CCR cake, and (b) dry CCR powder (author's own)

(a) (b)

Figure 7.4 SEM image of CCR powder (author's own)

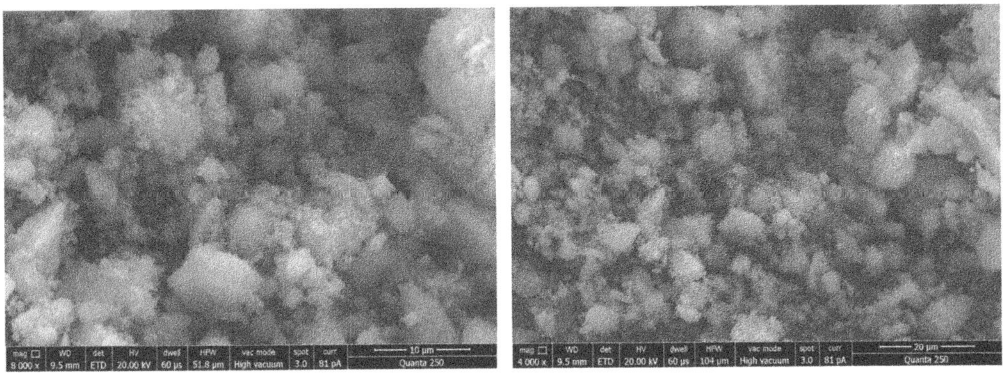

Figure 7.5 Powder XRD pattern of dry CCR (P = Portlandite, $Ca(OH)_2$; C = Calcite, $CaCO_3$) (author's own)

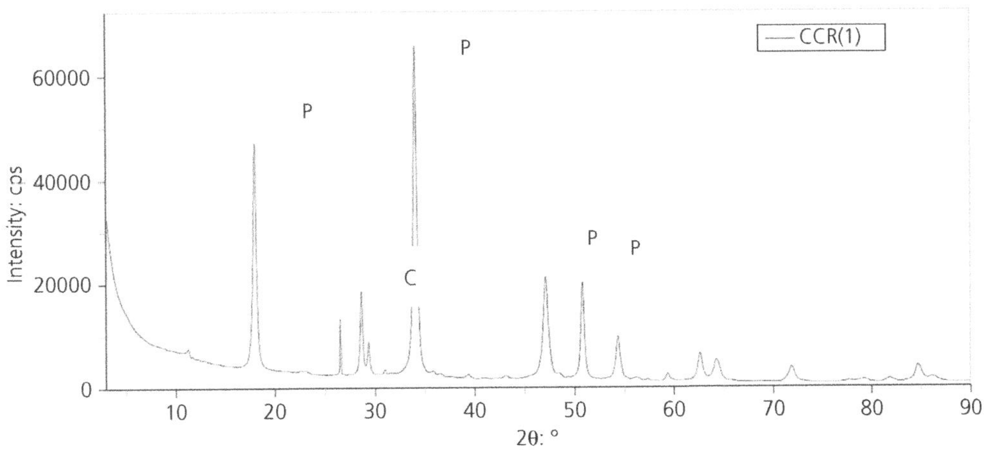

reported to reduce Portland cement consumption by 70% (Rattanashotinunt *et al*., 2013). Synergistic applications such as SCM and accelerated carbonation curing using CCR in concrete were investigated by Bawab *et al*. (2024). Enhanced compressive strength and CO_2 uptake resulting in densified microstructure as well as reduced CO_2 footprint were found in this research for replacing cement with CCR.

Olivine

Olivine ($(Mg, Fe)_2SiO_4$) is a natural rock-forming mineral abundant on the earth's surface found in igneous and metamorphic rocks. Around 3.3 Mt of olivine is consumed annually in Europe, most of which is supplied by Norway. Other places like Italy, Japan, Mexico, Pakistan, Spain, and the USA also have deposits of olivine (NSW Government Resources, 2024). Although the crystalline nature and limited pozzolanic characteristics hindered its application as SCM, the high overall content of silica (30–40%) and magnesium (40–60%) in olivine has attracted interest due to the recovery of amorphous silica and/or nano-silica by carbon sequestration (Stopic *et al.*, 2019; Wang *et al.*, 2019) and the acid digestion process (Shah and Scott, 2022; Shanks *et al.*, 2024) for its potential use as SCM in concrete. A similar performance of nano-silica synthesised from olivine as an alternative to commercial water glass for its use as an activator for geopolymer concrete with a reduced carbon footprint was reported (Gao *et al.*, 2017).

Carbon nanomaterials (graphene)

The breakthrough of nanotechnology in materials science and the extraordinary physical and chemical properties of carbonaceous fibres have resulted in enhanced performance and durability of normal and high-strength concrete. When at least one of its dimensions is less than 100 nm, a material is termed a nanomaterial. Although incorporating reinforcement in concrete as bar or fibre form is an established method to control cracking at macro and micro levels, controlling cracking at the nanoscale is crucial for enhancing concrete durability and strength. Graphene, which is a two-dimensional nanosheet, is one of the most well-known allotropes of carbon. Using graphene in composite concrete with the aim of reducing the volume of concrete and/or cement content has been recommended in the ICE *Low Carbon Concrete Routemap* (Low Carbon Concrete Group, 2022). The other best-known allotropes of carbon are graphite, diamond, carbon nanotubes (CNT), and fullerene. The high mechanical strength, up to 130 GPa tensile strength, and 1.1 TPa elastic modulus, together with the valuable flexibility to accommodate deformation (Xu and Gao, 2015), make graphene fibres a potential reinforcement material. Moreover, the ability to offer various potential intelligent and functional properties for embedding carbon fibres in concrete has been reported, such as electrical conductivity, pressure sensitivity, temperature sensitivity, electromagnetic wave shielding, and wave absorption (Tian *et al.*, 2021; Wang *et al.*, 2008; Wu *et al.*, 2023). The potential attractive application regarding structural durability and strength of concrete using nanomaterials (graphene oxide and nanosheets) was reviewed and reported by Shamsaei *et al.* (2018), and interest in experimental studies on the nanoscale interaction between graphene-based nanomaterials and cement matrices is growing.

Although carbon-based nanomaterials can be incorporated as admixtures or SCMs in concrete, achieving their appropriate and effective dispersion given their nanoscale size with ultra-high SSA is critical and presents significant challenges in realising their full potential benefits. Various dispersion techniques such as mechanical, ultrasonic, acid treatment, surface modification, silica fume, and so on have been used by researchers.

The higher reactivity and superior dispersion within cementitious systems due to the presence of hydroxyl, epoxide, carboxyl, and carbonyl functional groups on the surface of graphene oxide (GO) have made them attractive for their application in composite concrete (Compton and Nguyen, 2010). Graphene oxide-grafted carbon nanofibre (CNF-GO) was prepared by the 'grafting to' chemical graft method in a study by Wang *et al.* (2024b) and compared against plain and carbon nanofibre-modified concrete (CNFMC). The CNF-GO-modified concrete (CNF-GOMC) was found to provide the best dispersion effect, minimum resistivity, and maximum strength, while CNF-GO improved the strength and toughness of concrete compared to plain concrete and CNFMC due to improved interlocking force and additional formation of C–S–H crystal at the CNF-GO/concrete matrix interface.

Due to differences in size and orientation at the microstructural level, the influence on the mechanical strength of concrete is different for different types of carbon nanomaterials. Generally, adding GO is more effective than CNT or CNF for compressive strength, whereas CNT and

CNF were found to be more effective for improving the flexural and tensile strengths of concrete (Lu *et al.*, 2022).

High-strength concrete (HSC)

The Concrete Society (2024) defines high-strength concrete as having a specified characteristic cube strength between 60 and 100 MPa, although strengths in the range of 80–100 MPa are commonly used elsewhere. With technological advancements and innovations, the strength capacity of concrete continues to rise, and UHSC or reactive powder concrete have been classified with strengths reaching up to 500 N/mm² (Aïtcin and Lessard, 2019). High-strength (C65/80) concrete was used in the 300 m-high Shard – the tallest building in Western Europe – whereas concrete with 80 MPa characteristic cube strength was utilised in the 452 m-high Petronas Towers, Kuala Lumpur (Viswanath *et al.*, 1998). In 2007, an 80 MPa high-performance concrete mix to withstand the harsh local climate, with high chloride and high sulfate ground conditions, was used in the Burj Khalifa, Dubai, which is the world's tallest building (Elkem, 2023). The benefits in terms of overall structural economies such as reduced section size and materials consumption, and associated savings in space as well as higher span length with fewer supports make the use of HSC popular in addition to its high performance and durability features.

To manufacture HSC, various mix design parameters such as the use of: a low w/b ratio as low as 0.2; reactive SCM with ultrafine particle size such as silica fume; compatible superplasticisers; appropriately graded (for particle packing) high-quality aggregates with a maximum 10 mm size; as well as extensive trials are recommended (Soutsos et al., 2017). HSC is susceptible to self-desiccation, and the formation of micro-cracking for reduced w/b ratio implies the importance of curing (Persson, 1997). Avoidance of dry curing/air curing and low temperature (Sidhu and Siddique, 2024) as well as the application of steam curing at high temperatures for the formation of C–S–H with improved microstructure for pozzolanic reaction between C–H and nano-silica or silica fume for UHSC was also suggested (Janković *et al.*, 2019). Prestressed UHSC panels with dimensions of 1000 x 2000 x (75–100) mm, featuring a compressive strength of 164.2 MPa, tensile strength of 22.28 MPa, and elastic modulus of 44 GPa, demonstrated resilience under a 6 trinitrotoluene equivalent explosion (Ngo *et al.*, 2007). This performance suggests suitability for applications where blast and impact loading are critical considerations.

The use of synthetic SCMs or additives such as graphene which may serve to reduce the embodied carbon of concrete through reduced materials consumption and cement content has been suggested in the ICE *Low Carbon Concrete Routemap* (Low Carbon Concrete Group, 2022). A well-cured, properly designed concrete with a low water-to-binder (w/b) ratio, which is dense and less porous, is less susceptible to degradation, ensuring high performance regardless of its strength. Consequently, high-performance concrete (HPC) does not necessarily imply high strength, and similarly, high-strength concrete may not always exhibit high performance.

High-performance concrete (HPC)

While concrete is a durable and versatile material, its structural and aesthetic performance can be significantly compromised by various environmental and structural loads. If durability is not prioritised from the conceptual design stage through design, detailing, specification, and construction, and if planned inspection and maintenance are neglected, even HSC can deteriorate prematurely, potentially resulting in demolition and reconstruction well before its intended design life. In particular, in highly aggressive environments such as those with repeated wetting and drying, repeated freeze–thaw cycles, contaminated soils and aggregates, and exposure to acidic chemical effluents, designing HPC rather than solely HSC is increasingly advisable. This approach is also endorsed by standards to address the unpredictable weather events associated with climate change, ensuring durability and resilience – for example, BS 8500-1 (BSI, 2021a), which complements BS EN 206 (BSI, 2021c) by providing methods for specifying concrete, suggests that the compressive strength of concrete is not a limiting factor against chloride attack; the recommended strength class of concrete has, therefore, been removed, leaving the maximum w/c ratio and minimum cement content for all types of chloride exposure conditions. Hence HPC should be defined as concrete with strength, abrasion resistance, and durability significantly above normal and high-strength concrete. Chapter 4 discusses how moisture and chemical ingress initiate nearly all types of concrete degradation where high elastic modulus and low thermal strain and creep are essential to counteract these degradation phenomena. Accordingly, Mehta and Aïtcin (1990) suggest that impermeability and dimensional stability should be key criteria for qualifying as HPC.

Designing HPC requires a stringent, detailed approach that is tailored to the specific environmental conditions the concrete will face and expected degradation mechanisms. The common practices include using high cement content, high-quality aggregates, a low water/binder ratio (0.2–0.3), highly reactive superfine pozzolans, and superplasticisers to achieve minimal permeability and optimal compressive strength. More binder grains with low water per unit volume are supportive to produce dense microstructure in addition to a higher amount of C–S–H and dense ITZ in HPC. Although, this approach can increase costs by 30–100% over standard concrete, as noted by Leung (2001), reduced maintenance with enhanced service life make HPC more sustainable compared to NSC (Yang *et al.*, 2019).

Seismic design and increasing demand for skyscrapers, long-span bridges with heavy traffic, tunnelling, and offshore structures, as well as technological advancements in materials science (nanoparticles) and chemistry, have enabled the development of UHPC, which is defined by its specialised composition and exceptional properties. UHPC typically consists of fine-grained sand (eliminating coarse aggregate for optimised packing density), silica fume, steel fibres, and custom blends of high-strength Portland cement. This composition results in compressive strengths around 150 MPa and tensile strengths reaching 15 MPa, delivering enhanced durability and ductility that distinguish UHPC from conventional concretes (Wu *et al.*, 2018). To improve the ductility performance of unreinforced UHPC, use of different steel, glass, polymer, and carbon fibres are in use where dosages, size, and type of fibres determine the bonding within a given UHPC. Possessing a higher modulus of elasticity than the concrete matrices, steel fibres were found to enhance load-carrying capacity and impact, shatter, and abrasion resistance among different types of fibres (Wu *et al.*, 2018). Application of various nanomaterials such as carbon nanotubes, nano-SiO_2, nano-$CaCO_3$, and so on were found to provide accelerated hydration, shortened setting time, and increased early- and long-term strength in addition to enhancing the microstructure of cement composites at the nanoscale in UHPC.

Despite its potential sustainability benefits, several challenges limit the widespread application of UHPC in construction. Following an extensive review of UHPC technology, Amran *et al.* (2022) highlighted the following key obstacles.

- Thermal curing requirements restrict onsite applications and add to costs.
- High-cost and energy-intensive materials are often required.
- Complex production processes and limited expertise in the construction sector hinder adoption.

- Inefficient mixing methods for nanomaterials restrict their effective incorporation in mixes with higher dosages.
- The absence of standardised guidelines complicates implementation.
- Critical future availability of currently used pozzolans and SCMs raises sustainability concerns.

Self-compacting concrete (SCC)

BS EN 206 (BSI, 2021c) defines self-compacting concrete as 'concrete that is able to flow and compact under its own weight, fill the formwork with its reinforcement, ducts, boxouts, etc., whilst maintaining homogeneity'. As the name suggests, SCC achieves compaction under its weight without requiring mechanical effort. Its advantages over normal concrete in various demanding situations include easier placement in congested reinforcement zones, suitability for concreting in narrow trenches, efficient repair or strengthening of existing structures, high-rise buildings, and in situ pile construction. Additionally, SCC requires fewer workers for placement, making it an increasingly popular choice in modern construction.

Recently, the growing practice of using narrow and micro-trenching in utility sectors has attracted widespread application of SCC where the usual compacting effort (roller, plate compactor) cannot be used. Slump flow, viscosity, passing ability, and segregation resistance parameters are used to characterise SCC, implying the requirement of high fluidity and stability of the mixes. The slump-flow value for SCC lies in the range of 550–850 mm measured by BS EN 12350-8 (BSI, 2019b). A typical SCC mix proportion containing 28–34% of concrete volume taken by coarse aggregate (for normal concrete it is in the range of 40–55%), water/binder ratio in the range of 0.3–0.4 with 450–600 kg/m^3 binder content is suggested (Soutsos and Domone, 2017). As this mixture is unable to support itself, considerably stiffened formwork is usually required to withstand hydrostatic pressure exerted by the plastic SCC.

The influence of coarse aggregate gradation size 5–10 mm coarse aggregate weight/size 10–20 mm coarse aggregate weight (A/B ratio) in the range of 4/6 to 7/3 on SCC was examined by Zhao *et al.* (2012), who found reduced flow and increased wet density for increasing ratio attributed to higher water absorption and higher density of 5–10 mm coarse aggregate, respectively. Furthermore, maximum particle bulk density and less porosity for SCC with an A/B ratio of 4/6 were reported to show the highest compressive strength in this research. Concrete mixes that combine high strength with self-compacting capabilities hold significant potential for applications in structurally and environmentally demanding conditions. A study (Somasri and Narendra Kumar, 2021) reported progressive improvements in both slump flow and mechanical strength with increasing GO content in a ternary blended high-strength self-compacting concrete (HSSCC), where 33% of the cement was replaced by ground granulated blast furnace slag (GGBS, 30%) and silica fume (3%). After 28 days, the HSSCC incorporating 0.1% GO achieved compressive and flexural strengths of 73.6 MPa and 10.41 MPa, respectively, compared to 62.7 MPa and 8.4 MPa in the control mix containing no GO. Similar to traditional concrete, self-compacting geopolymer mortar and concrete are gaining significant interest. Variations in binders and activator concentrations have been reported to significantly influence early strength development, highlighting the importance of mix optimisation for enhanced performance, especially for emergency repair works (Thakkar *et al.*, 2023).

Artificial intelligence (AI) in concrete technology

In the era of Industry 4.0 and modern methods of construction, the integration of AI is gaining significant traction across the construction sector. AI is being applied to enhance safety, efficiency, and accuracy in research, development, design, bidding, transportation, real-time site monitoring, inspections, early warnings, forecasting, quality control (QC)/quality assurance (QA), and so on. By mimicking human intelligence, AI performs complex tasks using machine learning (ML), a subset of AI. ML employs algorithms further refined through deep learning (DL), which leverages artificial neural networks to solve intricate problems and optimise decision-making processes. The trial-and-error approach commonly used in concrete research heavily depends on the researcher's expertise, making it both resource-intensive and time-consuming – for example, petrographic image analysis at the microstructural level is a critical tool in materials processing and concrete technology, with SEM being a widely adopted technique. This process typically involves sample preparation steps such as impregnation, polishing, and coating, followed by image capture at various magnifications. However, the traditional method of data collection and analysis remains semi-automated, inflexible, labour-intensive, time-consuming, and prone to lower accuracy (Yazdi and Sarafrazi, 2014). Similarly, traditional visual condition surveys and semi-invasive sample extractions used for structural health monitoring (SHM) in asset management heavily rely on the engineer's expertise. This approach is not only time-consuming but also often fails to provide comprehensive and representative data, limiting its effectiveness in accurately assessing structural integrity.

In this case, a data-driven ML approach is expected to solve the complex, multidimensional, and in many cases nonlinear problems associated with the trial-and-error-based approach. The images obtained using SEM imaging can be used to train the DL algorithms to segment different components of the SEM image, which is crucial to understanding images at various granularity levels. Using semantic segmentation, a U-Net architecture-based convolutional neural network (CNN) model was proposed for image segmentation with a limited dataset (Bangaru *et al.*, 2022), where the model was developed, trained, and evaluated. This technique was found to automatically segment the SEM images of concrete into four classes: 1) aggregates; 2) anhydrous cement; 3) hydrates; and 4) pores, at the simplest heterogeneity level with an overall accuracy of 94.43%. Figure 7.6 illustrates actual, ground truth, and predicted images for different specimens and magnifications. A comprehensive review of AI-powered insights with image-centric approaches for concrete SHM, property exploration, and quality control was reported by Sarkar *et al.* (2024).

Substantial research efforts have focused on several AI-inspired non-destructive testing (NDT) technologies to improve the speed and accuracy of SHM and QC/QA methods and to enable the state of practice to evolve beyond destructive testing. NDT methods include acoustic emission signals for damage analysis (Carnì *et al.*, 2017); electromagnetic-wave (EM-wave) spectrum (Putranto *et al.*, 2024) for assessing micro-cracks of concrete structures in contact with water; ground penetrating radar for crack detection (Hussein *et al.*, 2024); and laser-induced breakdown spectroscopy for corrosion detection (Wudil *et al.*, 2024) where AI-inspired ML and DL algorithms with data collected using NDT were applied to predict the desired condition in a non-destructive way.

In the case of design for manufacture and assembly, offsite construction, building information modelling, unmanned aerial vehicles, 3D concrete, and so on, the application of construction robot process automation is essential where AI plays the most important role for Industry 5.0 transformation in construction.

Figure 7.6 Performance of U-Net CNN model in SEM image segmentation of concrete sample (Bangaru *et al.*, 2022)

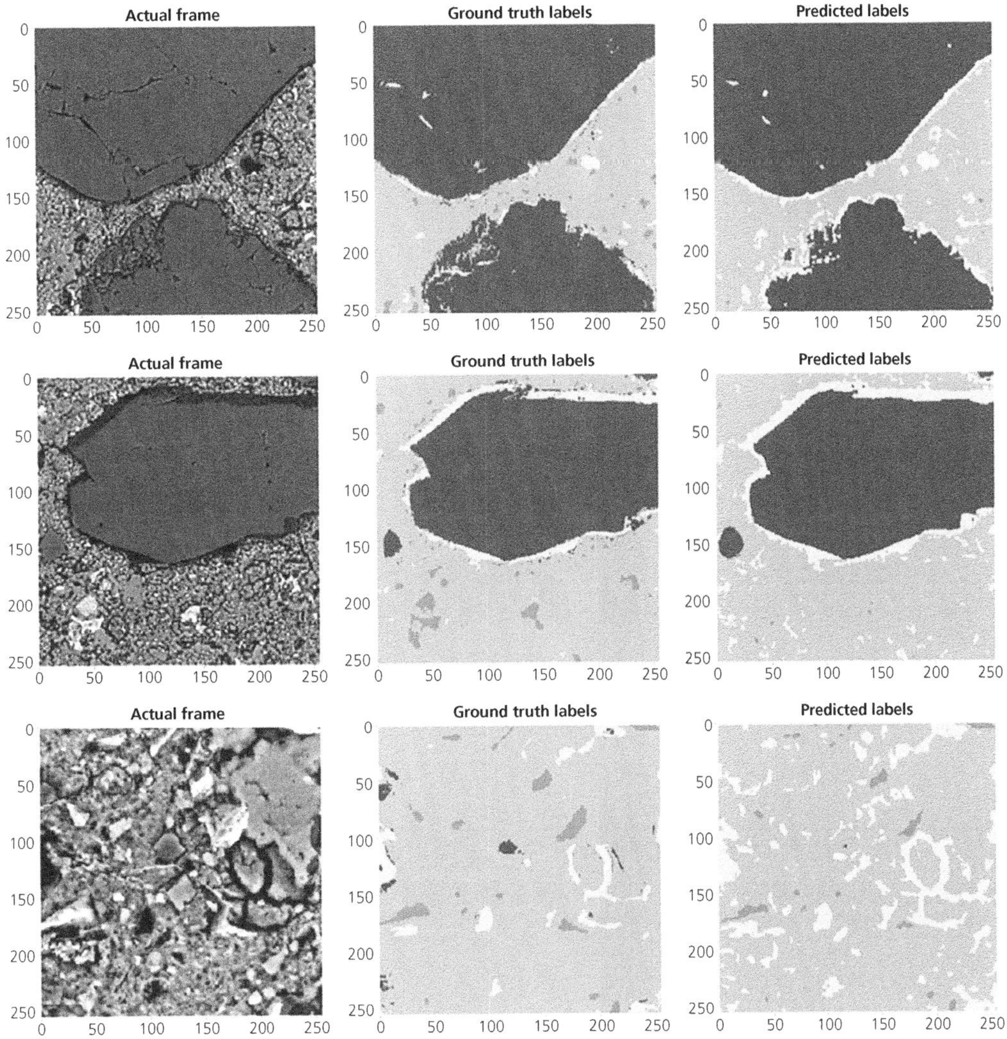

Monower Sadique
ISBN 978-1-83608-477-8
https://doi.org/10.1108/978-1-83608-474-720251008

Appendix: Solutions to questions for practice

Chapter 1

Question 1.1

Four major elements that are present in Portland clinker are calcium, silicon, aluminium, and iron. All these elements in Portland cement are present as compounds or phases. The four major phases, in shorthand form, are tricalcium silicate (C_3S) or alite, dicalcium silicate (C_2S) or belite, tricalcium aluminate (C_3A) or aluminate, and tetracalcium aluminoferrite (C_4AF) or ferrite.

Question 1.2

Portland cement is made by burning quarried natural limestone ($CaCO_3$) to a temperature of around 1500°C and nodules of clinker are produced through partial fusion. During this calcination, the carbonate component of limestone, which is a sedimentary rock, is converted to calcium oxide (CaO) and produces calcium silicates (which are the fundamental component of Portland cement), in addition to releasing carbon dioxide (CO_2) through the calcining reaction. The cooled clinker is then pulverised to fine grey powder to produce cement.

Question 1.3

As the content of clinker is minor (C) using the nomenclature, this type of cement can be designated as blast furnace cement BS EN 197-CEM III/C 32.5 L

Question 1.4

Concrete sectoral carbon emission can be reduced by contributing efficiency in design and construction, efficiency in concrete production, savings in cement and binders, savings in clinker productions, carbon capture, utilisation and storage at cement plant, CO_2 sink/recarbonation, decarbonisation of electricity, and so on.

Chapter 3

Question 3.1

The hydration of tricalcium silicate (C_3S) as well as dicalcium silicate (C_2S) after 3 to 12 hours accelerates and during this stage C-S-H starts to be formed. The concentration of Ca^{2+} and SO_4^{2-} ions in the liquid phase gradually declines due to the formation of portlandite (C-H) and ettringite, respectively. The adsorption of SO_4^{2-} ions onto the surface of C-S-H also reduces the ion concentration of sulphates. This stage of hydration is also termed as the acceleration stage when most of the strength-generating agent C-S-H is formed due to the dilution of C_3S and C_2S.

Question 3.2

The ITZ is a thin region of cement/binder pastes around the aggregates that govern bonding (chemical and mechanical) and strength at the microscopic level. The ongoing hydration increases C-S-H content, reducing porosity and enhancing ITZ density. In normal strength concrete, however, the ITZ is often the weakest link, where micro-cracks can develop under loading. The aggregate size and water/cement ratio affect the thickness of the ITZ and its impact on the mechanical properties of concrete. Due to the lower overall surface area of larger aggregates, a weaker transition zone, with a more critical effect on the concrete's strength at a lower water/cement ratio, is suggested.

Question 3.3

Concrete compressive strength is inversely proportional to the water/cement ratio. At very high water/cement ratios, the cement paste becomes highly porous, reducing the strength of the concrete. On the other hand, at very low water/cement ratios, the workability of concrete drastically decreases, making it difficult to achieve complete compaction. This results in concrete with higher void content and lower strength. At the early stages of hydration with a higher water-to-binder ratio, the microstructure develops with more ettringite and portlandite than C-S-H due to the dissolution of C_3A. This negatively affects the bond and porosity at the ITZ between hardened cement paste (HCP) and aggregate, thus reducing the durability of concrete.

Question 3.4

According to Abrams' law, the strength of concrete is inversely proportional to the water/binder (w/b) ratio once the minimum amount of water required to partially complete the hydration reaction is met. While a higher w/b ratio improves the workability of concrete, it also increases porosity, which in turn reduces strength and durability. Conversely, a lower w/b ratio enhances strength, reduces porosity, and improves durability but can result in reduced workability, potentially leading to incomplete compaction and defects.

The use of water-reducing admixtures, including plasticisers and superplasticisers, addresses this trade-off by reducing the amount of water needed to achieve a given workability. These admixtures allow for either maintaining the same workability (slump) with a reduced water content or increasing workability at a constant water content. This results in concrete with higher strength, improved durability, and reduced porosity without compromising the ease of handling, placing, and compacting the mix.

Question 3.5

For large concrete pours, susceptibility to the creation of cold joints between two successive layers is very likely. In that case, a retarder admixture that prolongs the setting time of cementitious binders by slowing down the hydration reaction in the initial hours is required. This also extends the workability needed for long hauls. Retarders work by adsorbing onto the binder particles, reducing their solubility and slowing the hydration process.

Chapter 4

Question 4.1

In this case, it is highly likely that the soil may contain sulphates as fertiliser was used previously for agriculture. Sodium, potassium, calcium, and magnesium sulphates are all common, and when in solution these will attack concrete. We can see that this problem mainly occurs in concrete in contact with the ground e.g. in foundations. In this type of attack, sulphates and water react with the tricalcium aluminate found in Portland cement to form a calcium sulfo-aluminate hydrate ($3CaO.Al_2O_3.CaSO_4.31H_2O$) known as ettringite. In the presence of water, sulphate ions can react with tricalcium aluminate compound (C_3A). This reaction causes expansion, leading to cracking and eventual disintegration of the concrete.

The following factors need to be considered while designing concrete for this foundation

- design concrete with low permeability
- correct choice of materials for concrete – use GGBS
- cement type (sulphate resisting) – low C_3A
- protective coatings.

Question 4.2

Steel reinforcement in concrete is protected from corrosion primarily by the alkaline environment provided by hydrated cement, which typically has a pH of 12.6 or higher. This high pH environment promotes the formation of a passive oxide film on the surface of the steel, significantly reducing the rate of corrosion to a negligible and non-damaging level. This protective mechanism remains effective as long as the concrete cover is intact and not exposed to aggressive chemicals or physical damage. Therefore, the concrete cover offers both chemical protection, through alkalinity, and physical protection by acting as a barrier to external contaminants.

Question 4.3

Following the relationship between carbonation depth (x in mm) and time (t in year) or duration of exposure

$$20 = k\sqrt{16}$$

$$k = 5$$

$40 = 5\sqrt{t}$

$t = 64$ years

Thus, the carbonation front is expected to reach the reinforcement after 64 years, or 48 years from now.

Question 4.4

In achieving this, a well-cured and well-compacted good quality concrete containing additions (SCM) with low permeability and sufficient clear cover is essential. Accordingly, the standard specified threshold values for the following mix design parameter are

- most severe exposure class (permanent or wet, rarely dry or wet, cyclic wet and dry)
- maintaining concrete strength class based on exposure (minimum)
- maintaining water/cement ratio (maximum)
- maintaining cement content (minimum)
- maintaining cover to reinforcement (minimum)
- correct cement type.

Question 4.5

During exposure to fire, cement paste undergoes dehydration and shrink, whereas aggregates expand at high temperatures. At around 500°C, a number of physical and chemical changes occur within concrete in addition to induced thermal stress due to disparity in thermal expansion between the HCP and the aggregate, particularly at the ITZ, leading to cracking and rapid strength loss. The mismatch in thermal expansion becomes even more pronounced above 600°C when coarse aggregates continue to expand while the C-S-H gel shrinks due to the loss of both physically and chemically bound moisture. This opposing physical response at different temperatures creates internal stresses and explains why a low aggregate/cement ratio is desirable. The low aggregate/cement ratio will minimise the opposing internal stress by reducing the proportion of aggregate relative to cement paste.

Question 4.6

Once the passive protective layer around steel reinforcement within concrete is compromised, corrosion can occur provided two conditions are met: (1) the presence of sufficient moisture and (2) adequate oxygen. Interestingly, these two requirements tend to counteract each other. When the concrete is saturated, oxygen penetration is restricted, limiting corrosion. Conversely, when the concrete is dry, while oxygen availability is high, the lack of moisture hinders the corrosion process. However, in structural members exposed to repeated wetting and drying cycles, both moisture and oxygen are intermittently available, creating an ideal environment for accelerated corrosion. This cyclic exposure significantly increases the risk of steel reinforcement degradation.

Question 4.7

The ASR occurs when the alkalis present in cement react with specific forms of silica in aggregates, resulting in the formation of a hygroscopic silica gel. For this reaction to take place, both reactive silica and alkalis must be present. Reactive silica is often found in certain aggregates, such as contaminated flints, limestones, cherts, and tuffs. Alkalis are primarily supplied by cement with a high alkali content. The reaction produces a gel that can weaken the bond between the aggregates and the HCP. This gel absorbs water, swells, and generates internal stresses, leading to characteristic map-like cracking and the eventual deterioration of the concrete.

The controlling measures include

(*a*) using aggregates which experience has shown to perform satisfactorily
(*b*) using cement with a low alkali content
(*c*) inhibiting the ingress of water.

Chapter 5

Question 5.1

Hydraulic and pozzolanic are two primary categories of SCMs. Hydraulic materials, such as Portland cement, react directly with water to form cementitious compounds, while pozzolanic materials require calcium hydroxide (C-H) and water to form cementitious products. Fly ash is exclusively a pozzolanic material with no inherent hydraulic properties, whereas GGBS exhibits both latent hydraulic and pozzolanic characteristics.

When a high proportion (e.g. 70%) of Portland clinker is replaced with fly ash, the limited availability of C-H can hinder the formation of sufficient cementitious compounds, compromising strength development. In contrast, GGBS, due to its partial hydraulic reactivity, can independently form cementitious products even when C-H is limited, enabling higher replacement levels without significantly compromising the performance of the concrete.

Question 5.2

Using the principles of cement designation, CEM II/B-V contains a higher proportion of fly ash and correspondingly less Portland clinker compared to CEM II/A-V. Since clinker production is the primary source of embodied carbon in cement, the reduced clinker content in CEM II/B-V results in significantly lower embodied carbon.

The higher proportion of SCMs like fly ash in CEM II/B-V leads to slower strength development compared to cement with higher clinker content. This slower strength gain may not impact the construction schedule in scenarios where early strength is not critical, such as in foundations that are cast against the ground and subjected to gradually increasing loads.

Although standard concrete strength is measured at 28 days, concrete made with a higher proportion of fly ash continues to gain strength beyond this period – for example, the 56-day strength of such concrete can be approximately 40% higher than its 28-day strength. Designers can capitalise on this characteristic by specifying the required strength at 56 days instead of 28 days. While this approach may necessitate slight modifications to the design, it can lead to substantial reductions in embodied carbon, contributing to more sustainable construction practices.

Question 5.3

SF is a by-product of silicon metal and ferro-silicon alloys in electric arc furnaces. Its extreme fineness (100 times finer than cement) also helps achieve higher pozzolanicity with other cementitious materials. SF is more reactive than other supplementary materials such as fly ash and GGBS. SF functions in concrete as a highly efficient pozzolan – that is, it reacts chemically with the calcium hydroxide produced by the hydration of the Portland cement to form C-S-H which bind the concrete together. Unlike GGBS and PFA, SF contains highly reactive amorphous silicon dioxide that enhances the dissolution of the C_3S component of cement, generating portlandite (C-H) and C-S-H gel compounds. This process not only densifies the concrete complemented by the filler effect of SF's superfine particles but also maintains early strength similar to reference concrete. However, due to their high SSA, the water demand for concrete increases. HRWRA are commonly used to adjust the water content to attain the desired w/c ratio and workability of the concrete.

Monower Sadique
ISBN 978-1-83608-477-8
https://doi.org/10.1108/978-1-83608-474-720251011

References

Abd-El.Aziz MA, Abd.El.Aleem S and Heikal M (2012) Physico-chemical and mechanical characteristics of pozzolanic cement pastes and mortars hydrated at different curing temperatures. *Construction and Building Materials* **26**: 310–316.

Abrams DA (1919) *Design of Concrete Mixtures*. Structural Materials Research Laboratory, Lewis Institute, Chicago, IL, USA.

Ahmad S, Baghabra Al-Amoudi OS, Khan SMS and Maslehuddin M (2022) Effect of silica fume inclusion on the strength, shrinkage, and durability characteristics of natural pozzolan-based cement concrete. *Case Studies in Construction Materials* **17**: e01255.

Aïtcin P-C and Lessard J-M (2019) The composition and design of high-strength concrete and ultrahigh-strength concrete. In *Developments in the Formulation and Reinforcement of Concrete*, 2nd edn (Mindess S (ed.)). Woodhead Publishing, Cambridge, UK.

Aïtcin P-C (2019) The influence of the water/cement ratio on the sustainability of concrete. In *Lea's Chemistry of Cement and Concrete*, 5th edn (Hewlett PC and Liska M (eds)). Butterworth-Heinemann, Oxford, UK.

Akduman S, Aktepe R, Aldemir A *et al.* (2024) Opportunities and challenges in constructing a demountable precast building using C and D waste-based geopolymer concrete: a case study in Türkiye. *Journal of Cleaner Production* **434**: 139976.

Alabbasi M, Agkathidis A and Chen H (2023) Robotic 3D printing of concrete building components for residential buildings in Saudi Arabia. *Automation in Construction* **148**: 104751.

Al-Tabbaa A (2013) Reactive magnesia cement. In *Eco-Efficient Concrete* (Pacheco-Torgal F, Jalali S, Labrincha J and John VM (eds)). Woodhead Publishing, Cambridge, UK.

Ambroise J, Murat M and Péra J (1985) Hydration reaction and hardening of calcined clays and related minerals V. Extension of the research and general conclusions. *Cement and Concrete Research* **15**: 261–268.

Amran M, Huang S-S, Onaizi AM *et al.* (2022) Recent trends in ultra-high performance concrete (UHPC): Current status, challenges, and future prospects. *Construction and Building Materials* **352**: 129029.

Antiohos SK, Papadakis VG, Chaniotakis E and Tsimas S (2007) Improving the performance of ternary blended cements by mixing different types of fly ashes. *Cement and Concrete Research* **37**: 877–885.

Arjunan P, Silsbee MR and Roy DM (2001) Chemical activation of low calcium fly ash Part 1: Identification of suitable activators and their dosage. *2001 International Ash Utilisation Symposium*, Lexington, KU, USA.

Ashraf WB and Noor MA (2011) Performance-evaluation of concrete properties for different combined aggregate gradation approaches. *Procedia Engineering* **14**: 2627–2634.

Aydin S, Karatay Ç and Baradan B (2010) The effect of grinding process on mechanical properties and alkali–silica reaction resistance of fly ash incorporated cement mortars. *Powder Technology* **197**: 68–72.

Atkins C (2015) Physical deterioration mechanisms. In *ICE Handbook of Concrete Durability* (Soutsos M (ed.). Emerald, Leed, UK.

Azar P, Samson G, Deby F *et al.* (2023) Transfer properties and chloride-induced corrosion resistance of GGBS and metakaolin alkali-activated materials. *Cement and Concrete Composites* **142**: 105182.

Babu KG and Kumar VSR (2000) Efficiency of GGBS in concrete. *Cement and Concrete Research* **30**: 1031–1036.

Bai J, Chaipanich A, Kinuthia JM *et al.* (2003) Compressive strength and hydration of waste-paper sludge ash-ground granulated blastfurnace slag blended pastes. *Cement and Concrete Research* **33**: 1189–1202.

Banfill P and Frias M (2007) Rheology and conduction calorimetry of cement modified with calcined paper sludge. *Cement and Concrete Research* **37**: 184–190.

Bangaru SS, Wang C, Zhou X and Hassan M (2022) Scanning electron microscopy (SEM) image segmentation for microstructure analysis of concrete using U-net convolutional neural network. *Automation in Construction* **144**: 104602.

Bawab J, El-Hassan H, El-Dieb A and Khatib J (2024) Accelerated carbonation curing of concrete incorporating calcium carbide residue. *Journal of Building Engineering* **88**: 109258.

BBC (2024) Moment UK's last coal-fired power station shuts down. BBC, London, UK.

Beaudoin J and Odler I (2019) Hydration, setting and hardening of Portland cement. In *Lea's Chemistry of Cement and Concrete*, 5th edn (Hewlett PC and Liska M (eds)). Butterworth-Heinemann, Oxford, UK.

Ben-Bassat M, Nixon PJ and Hardcastlet J (1990) The effect of differences in the composition of Portland cement on the properties of hardened concrete. *Magazine of Concrete Research* **42**: 59–66.

Bentz D, Garboczi J and Stutzman P (1992) Computer modelling of the interfacial zone in concrete. *Rilem Proceedings*. Chapman and Hall, London, UK, p.107.

Bhanja S and Sengupta B (2003) Modified water–cement ratio law for silica fume concretes. *Cement and Concrete Research* **33**: 447–450.

Biomass Energy Centre (2012) *What is Biomass*? http://www.biomassenergycentre.org.uk/portal/page?_pageid=76,15049&_dad=portal&_schema=Portal (accessed 16/03/2012).

Blanco F, Garcia MP and Ayala J (2005) Variation in fly ash properties with milling and acid leaching. *Fuel* **84**: 89–96.

Blotevogel S, Doussang L, Poirier M *et al.* (2024) The influence of Al_2O_3, CaO, MgO and TiO_2 content on the early-age reactivity of GGBS in blended cements, alkali-activated materials and supersulfated cements. *Cement and Concrete Research* **178**: 107439.

Botha A and Strydom CA (2001) Preparation of a magnesium hydroxy carbonate from magnesium hydroxide. *Hydrometallurgy* **62**: 175–183.

Bouaissi A, Li L-Y, Al Bakri Abdullah MM and Bui Q-B (2019) Mechanical properties and microstructure analysis of FA-GGBS-HMNS based geopolymer concrete. *Construction and Building Materials* **210**: 198–209.

Bouzoubaâ N, Zhang MH, Bilodeau A and Malhotra VM (1997) The effect of grinding on the physical properties of fly ashes and a Portland cement clinker. *Cement and Concrete Research* **27**: 1861–1874.

BRE (Building Research Establishment) (1997) *Design of Normal Concrete Mixes*, 2nd edn. (Teychenne DC, Franklin RE and Erntroy HC (eds)). Building Research Establishment, Garston, UK.

BRE (2008) *Sulfate Damage to Concrete Floors on Sulfate-Bearing Hardcore – Identification and Remediation* (Department for Communities and Local Government (ed.)). RIBA Publishing, London, UK.

Britpave (British Cementitious Paving Association) (2022) *Guide to Roller Compacted Concrete Pavements Guidelines for Design, Materials and Construction*, 2nd edn. Britpave, Wokingham, UK.

Brooks JJ and Megat Johari MA (2001) Effect of metakaolin on creep and shrinkage of concrete. *Cement and Concrete Composites* **23**: 495–502.

BSI (2002) BS EN 12620:2002+A1:2008 Aggregates for concrete. BSI, London, UK.

BSI (2005a) BS EN 14647:2005 Calcium aluminate cement – composition, specifications and conformity criteria. BSI, London, UK.

BSI (2005b) BS EN 196-2:2005 Methods of testing cement – Part 2: Chemical analysis of cement. BSI, London, UK.

BSI (2006) BS EN 15167-1:2006 Ground granulated blast furnace slag for use in concrete, mortar and grout – Part 1: definitions, specifications and conformity criteria. BSI, London, UK.

BSI (2009a) BS 476-10:2009 Fire tests on building materials and structures – Part 10: guide to the principles, selection, role and application of fire testing and their outputs. BSI, London, UK.

BSI (2009b) BS EN 13263-1:2005+A1:2009 Silica fume for concrete – Part 1: Definitions, requirements and conformity criteria. BSI, London, UK.

BSI (2011a) BS EN 196-5:2011 Methods of testing cement – Part 5: pozzolanicity test for pozzolanic cement. BSI, London, UK.

BSI (2011b) BS EN 15978:2011 Sustainability of construction works – assessment of environmental performance of buildings – calculation method. BSI, London, UK.

BSI (2012a) BS EN 450-1:2012 Fly ash for concrete – Part 1: definition, specifications and conformity criteria. BSI, London, UK.

BSI (2012b) BS EN 934-2:2009+A1: 2012 Admixtures for concrete, mortar and grout. BSI, London, UK.

BSI (2016a) BS EN 196-3 Methods of testing cement – Part 3: Determination of setting times and soundness. Brussels: BSI, London, UK.

BSI (2016b) PAS 8820:2016 Construction materials. Alkali-activated cementitious material and concrete – specification. BSI, London, UK.

BSI (2018) BS EN 13501-1:2018 Fire classification of construction products and building elements – Part 1: Classification using data from reaction to fire tests. BSI, London, UK.

BSI (2019a) BS EN 197-1:2011 Cement – Part 1: composition, specifications and conformity criteria for common cements. BSI, London, UK.

BSI (2019b) BS EN 12350-8:2019 Testing fresh concrete. Self-compacting concrete. Slump-flow test. BSI, London, UK.

BSI (2019c) BS EN 12390-3:2019 Testing hardened concrete. Compressive strength of test specimens. BSI, London, UK.

BSI (2019d) EN 15804:2012+A2:2019 Sustainability of construction works – environmental product declarations – core rules for the product category of construction products. BSI, London, UK.

BSI (2020) BS EN 196-1:2016 – Methods of testing cement – Part 1: determination of strength. BSI, London, UK.

BSI (2021a) BS 8500-1:2023 Concrete – complementary British Standard to BS EN 206-Part 1: method of specifying and guidance for the specifier. BSI, London, UK.

BSI (2021b) BS 8500-2:2023 Concrete – complementary British Standard to BS EN 206-Part 2: specification for constituent materials and concrete. BSI, London, UK.

BSI (2021c) BS EN 206:2013+A2:2021 Concrete – Part 1: specification, performance, production and conformity. BSI, London, UK.

BSI (2021d) BS EN 197-5:2021 Cement – Part 5: Portland-composite cement CEM II/C-M and composite cement CEM VI. BSI, London, UK.

BSI (2022) BS EN 15804:2012+A2:2019 – Sustainability of construction works – environmental product declarations – core rules for the product category of construction products. BSI, London, UK.

BSI (2023) BS EN 13369:2023 Common rules for precast concrete products. BSI, London, UK.

BSI (2024) 2BSI Flex 350 v2.0:2024-09 Alternative binder systems for lower carbon concrete – code of practice. BSI, London, UK.

Buswell RA, Leal De Silva WR, Jones SZ and Dirrenberger J (2018) 3D printing using concrete extrusion: a roadmap for research. *Cement and Concrete Research* **112**: 37–49.

Cai M, Zhu, H, Wan Y *et al.* (2024) Using ultrafine fly ash to achieve low-carbon, high strength and high toughness engineered cementitious composites LC-HSTC. *Case Studies in Construction Materials* **20**: e03259.

Carnì DL, Scuro C, Lamonaca F, Olivito RS and Grimaldi D (2017) Damage analysis of concrete structures by means of acoustic emissions technique. *Composites Part B: Engineering* **115**: 79–86.

CCAA (Cement Concrete and Aggregates Australia) (2006) Curing of Concrete – Data Sheet. CCAA, West Perth, WA, Australia.

CCAA (2020) *Guide to Concrete Construction, v. 1. T41*. CCAA, West Perth, WA, Australia.

CEPI (Confederation of European Paper Industries) (2011) *CEPI Sustainability Report 2011*. CEPI, Brussels, Belgium.

CEPI (2024) *Key Statistics 2023*. CEPI, Brussels, Belgium.

Chatterjee AK (2001) X-Ray Diffraction. In *Handbook of Analytical Techniques in Concrete Science and Technology* (Ramachandran VS and James JB (eds)). William Andrew Publishing, Norwich, NY, USA.

Cheng A, Huang R, Wu J-K and Chen C-H (2005) Influence of GGBS on durability and corrosion behavior of reinforced concrete. *Materials Chemistry and Physics* **93**: 404–411.

Chindaprasirt P, Chotithanorm C, Cao HT and Sirivivatnanon V (2007) Influence of fly ash fineness on the chloride penetration of concrete. *Construction and Building Materials* **21**: 356–361.

Chung J, Lee G and Kim J-H (2021) Framework for technical specifications of 3D concrete printers. *Automation in Construction* **127**: 103732.

Chusilp N, Jaturapitakkul C and Kiattikomol K (2009) Utilization of bagasse ash as a pozzolanic material in concrete. *Construction and Building Materials* **23**: 3352–3358.

Collepardi M, Collepardi S, Skarp U and Troli R (2004) Optimization of silica fume, fly ash and amorphous nano-silica in superplasticized high-performance concrete. *Advancing Concrete Knowledge* **221**: 495–506.

Compton OC and Nguyen ST (2010) Graphene oxide, highly reduced graphene oxide, and graphene: versatile building blocks for carbon-based materials. *Small* **6**: 711–723.

Concrete Society Working Party (2000) *Diagnosis of Deterioration in Concrete Structures: Identification of Defects, Evaluation and Development of Remedial Action*. Technical Report No 54. Concrete Society, Hawley, UK.

Cordeiro GC, Toledo Filho RD, Tavares LM and Fairbairn EMR (2008) Pozzolanic activity and filler effect of sugar cane bagasse ash in Portland cement and lime mortars. *Cement and Concrete Composites* **30**: 410–418.

Cordeiro GC, Toledo Filho RD, Tavares LM and Fairbairn EMR (2012) Experimental characterization of binary and ternary blended-cement concretes containing ultrafine residual rice husk and sugar cane bagasse ashes. *Construction and Building Materials* **29**: 641–646.

CPI (Confederation of Paper Industries) (2021) Process Wastes from the Paper Industry. CPI, Swindon, UK.

Croft J (2024) Port Talbot: steel town braces for shutdown of its last blast furnace. *The Guardian*, September 30.

CSMA (Cementitious Slag Makers Association) (2024) *GGBS Concrete*. https://ukcsma.co.uk/ggbs-concrete/#:~:text=Typically%20the%207%2Dday%20strength%20of%20a%20ggbs,content%20concrete%2C%20kerb%20backing%20etc%2C%20the%20reduction (accessed 06/09/2024).

Davidovits J (1993) Geopolymer cements to minimize carbon dioxide greenhouse warming. *Ceramic Transactions*, **37**: 165–182.

Davidovits J (2008) *Geopolymer Chemistry and Applications*. Geopolymer Institute, Saint-Quentin, France.

Dransfield J (2009) Admixtures. In *ICE Manual of Construction Materials* (Forde M (ed.)). ICE, London, UK.

Dulaimi A, Al-Busaltan S and Sadique M (2021) The development of a novel, microwave assisted, half-warm mixed asphalt. *Construction and Building Materials* **301**: 124043.

Dulaimi A, Shanbara HK, Jafer H and Sadique M (2020) An evaluation of the performance of hot mix asphalt containing calcium carbide residue as a filler. *Construction and Building Materials* **261**: 119918.

Dung NT and Unluer C (2017) Sequestration of CO_2 in reactive MgO cement-based mixes with enhanced hydration mechanisms. *Construction and Building Materials* **143**: 71–82.

Dunster AM (2007) *Paper Sludge and Paper Sludge Ash in Portland Cement Manufacture*. Building Research Establishment (BRE), Watford, UK.

Dunster A (2009) *Silica Fume in Concrete*. Information Paper-IP 5/09. Building Research Establishment (BRE), Watford, UK.

Duxson P and Provis JL (2008) Designing precursors for geopolymer cements. *Journal of the American Ceramic Society* **91**: 3864–3869.

Edmeades RM and Hewlett PC (2003) Cement admixtures. In *Lea's Chemistry of Cement and Concrete*, 4th edn. Butterworth-Heinemann, Oxford, UK.

Elkem (2023) High strength concrete with Elkem Microsilica product. https://www.elkem.com/markets/smarter-cities-construction/strengthening-protective-applications/silica-fume-for-concrete (accessed 26/02/2025).

Esteves LP (2011) On the hydration of water-entrained cement–silica systems: combined SEM, XRD and thermal analysis in cement pastes. *Thermochimica Acta* **518**: 27–35.

Eymael MMT and Cornelissen HAW (1996) Processed pulverized fuel ash for high-performance concrete. *Waste Management* **16**: 237–242.

Fan YM, Yin SH, Wen ZY and Zhong JY (1999) Activation of fly ash and its effects on cement properties. *Cement and Concrete Research* **29**: 467–472.

Felekoglu B, Tosun K, Baradan B, Altun A and Uyulgan B (2006) The effect of fly ash and limestone fillers on the viscosity and compressive strength of self-compacting repair mortars. *Cement and Concrete Research* **36**: 1719–1726.

Fernandez-Jimenez A and Palomo A (2005) Composition and microstructure of alkali activated fly ash binder: effect of the activator. *Cement and Concrete Research* **35**: 1984–1992.

Ganganath N, Cheng CT, Fok KY and Tse CK (2016) Trajectory planning for 3D printing: a revisit to traveling salesman problem. *2016 2nd International Conference on Control, Automation and Robotics (ICCAR)*, 28–30 April 2016, pp. 287–290.

Gao X, Yu QL, Lazaro A and Brouwers HJH (2017) Investigation on a green olivine nano-silica source based activator in alkali activated slag-fly ash blends: reaction kinetics, gel structure and carbon footprint. *Cement and Concrete Research* **100**: 129–139.

Gartner EM and Macphee DE (2011) A physico-chemical basis for novel cementitious binders. *Cement and Concrete Research* **41**: 736–749.

Gaurav G, Kandpal SC, Mishra D and Kotoky N (2024) A comprehensive review on fly ash-based geopolymer: a pathway for sustainable future. *Journal of Sustainable Cement-Based Materials* **13**: 100–144.

GCCA (Global Cement and Concrete Association) (2021) *GCCA Concrete Future – Roadmap to Net Zero*. GCCA, London, UK.

GCCA (2023) *GCCA Concrete Future – Cement Industry Net Zero Progress Report*. GCCA, London, UK.

GCCA (2024a) *GCCA Policy Document on Blended Cements and Supplementary Cementitious Materials*. GCCA, London, UK.

GCCA (2024b) *Numerical Definitions for Low Carbon and Near Zero Emissions Concrete*. GCCA, London, UK.

Ghosal M (2023) Modern admixtures for durable and sustainable concrete. *Materials Today: Proceedings*.

Gollop RS and Taylor HFW (1994) Microstructural and microanalytical studies of sulfate attack. II. Sulfate-resisting Portland cement: ferrite composition and hydration chemistry. *Cement and Concrete Research* **24**: 1347–1358.

Grasserbauer M (1989) Materials analysis in high technology – From analytical chemistry to analytical science. *Trends in Analytical Chemistry* **8**: 191.

Gruskovnjak A, Lothenbach B, Winnefeld F *et al.* (2008) Hydration mechanisms of super sulphated slag cement. *Cement and Concrete Research* **38**: 983–992.

Hamed YR, Yousry Elshikh MM, Elshami AA, Matthana MHS and Youssf O (2024) Mechanical properties of fly ash and silica fume based geopolymer concrete made with magnetized water activator. *Construction and Building Materials* **411**: 134376.

Hänchen M, Prigiobbe V, Baciocchi R and Mazzotti M (2008) Precipitation in the Mg-carbonate system – effects of temperature and CO_2 pressure. *Chemical Engineering Science* **63**: 1012–1028.

Hanke LD (2001) *The Handbook of Analytical Methods for Materials*. Materials Evaluation and Engineering, Inc, Plymouth, MN, USA.

Hasani A and Dorafshan S (2024) Transforming construction? Evaluation of the state of structural 3D concrete printing in research and practice. *Construction and Building Materials* **438**: 137027.

Heah CY, Kamarudin H, Al-Bakri AMM *et al.* (2011) Effect of curing profile on kaolin-based geopolymers. *Physics Procedia* **22**: 305–311.

Heath A, Paine K, Goodhew S, Ramage M and Lawrence M (2013) The potential for using geopolymer concrete in the UK. *Proceedings of the Institution of Civil Engineers – Construction Materials* **166**: 195–203.

Heidrich C, Sanjayan J, Berndt M and Foster S (2015) Pathways and barriers for acceptance and usage of geopolymer concrete in mainstream construction. *Proceedings of the 2015 World of Coal Ash (WOCA), Nashville, TN, USA*.

Heikal M, Morsy MS and Aiad I (2006) Effect of polycarboxylate superplasticizer on hydration characteristics of cement pastes containing silica fume. *Ceramics-Silikaty* **50**: 5–14.

Higgins DD and Crammond NJ (2003) Resistance of concrete containing GGBS to the thaumasite form of sulfate attack. *Cement and Concrete Composites* **25**: 921–929.

Highways England (2020) *Highway Structures and Bridges Design: CD374 – The Use of Recycled Aggregates in Structural Concrete*. National Highways, Birmingham, UK.

Hojamberdiev M, Kameshima Y, Nakajima A, Okada K and Kadirova Z (2008) Preparation and sorption properties of materials from paper sludge. *Journal of Hazardous Materials* **151**: 710–719.

Holland TC and Detwiler R (2000) *Guide for the Use of Silica Fume in Concrete*. Report by ACI Committee 234. American Concrete Institute, Farmington Hills, MI, USA.

Hussein R, Etete B, Mahdi H and Al-Shukri H (2024) Detection and delineation of cracks and voids in concrete structures using the ground penetrating radar technique. *Journal of Applied Geophysics* **226**: 105379.

IEA (International Energy Agency) (2022) *Achieving Net Zero Heavy Industry Sectors in G7 Members*. IEA, Paris, France.

Irum S and Shabbir F (2024) Performance of fly ash/GGBFS based geopolymer concrete with recycled fine and coarse aggregates at hot and ambient curing. *Journal of Building Engineering* **95**: 110148.

Janković K, Bojović D and Stojanović M (2019) Influence of nanoparticles on the strength of ultra-high performance concrete. In *Nanotechnology in Eco-efficient Construction*, 2nd edn (Pacheco-Torgal F, Diamanti MV, Nazari A, Granqvist CG, Pruna A and Amirkhanian S (eds)). Woodhead Publishing, Cambridge, UK.

Jones C (2019) *Inventory of Carbon & Energy (ICE) Database – Embodied Carbon Model of Cement, Mortar and Concrete, v. 1.1 Beta, 28 Nov 2019*. Circular Ecology, London, UK.

Jones MR, Dhir RK, Csetenyi LJ and Csetenyi E (2004) *Alkali Pre-activation of PFA to Maximise Its Use in Concrete Construction*. University of Dundee, Dundee, Scotland.

Jones MR, Mccarthy A and Booth APPG (2006) Characteristics of the ultrafine component of fly ash. *Fuel* **85**: 2250–2259.

Joshi T and Dave U (2017) Parametric study of pervious concrete with changing cement content and w/c ratio. *Indian Concrete Journal* **91**: 19–25.

Kadhim A, Sadique M, Al-Mufti R and Hashim K (2020) Long-term performance of novel high-calcium one-part alkali-activated cement developed from thermally activated lime kiln dust. *Journal of Building Engineering* **32**: 101766.

Kadhim A, Sadique M, Al-Mufti R and Hashim K (2021) Developing one-part alkali-activated metakaolin/natural pozzolan binders using lime waste. *Advances in Cement Research* **33**: 342–356.

Kanagaraj B, Anand N, Raj RS and Lubloy E (2024) Behavioural studies on binary blended high strength self compacting geopolymer concrete exposed to standard fire temperature. *Ain Shams Engineering Journal* **15**: 102394.

Kang S, Wang J, Hu C *et al.* (2024) The influence of mechanochemical activation on the rheological properties and strength development of geopolymer. *Journal of Building Engineering* **96**: 110585.

Keleş ÖF and Akpinar MV (2022) Strength properties of roller compacted concrete pavement (RCCP) under different curing methods. *Construction and Building Materials* **324**: 126530.

Khoshnevis B (2004) Automated construction by contour crafting – related robotics and information technologies. *Automation in Construction* **13**: 5–19.

Khoury GA (1992) Compressive strength of concrete at high temperatures: a reassessment. *Magazine of Concrete Research* **44**: 291–309.

Kizhakkumodom Venkatanarayanan H and Rangaraju PR (2015) Effect of grinding of low-carbon rice husk ash on the microstructure and performance properties of blended cement concrete. *Cement and Concrete Composites* **55**: 348–363.

Kore Sudarshan D and Vyas AK (2019) Impact of fire on mechanical properties of concrete containing marble waste. *Journal of King Saud University – Engineering Sciences* **31**: 42–51.

Kumar S, Kumar R, Bandopadhyay A *et al.* (2008) Mechanical activation of granulated blast furnace slag and its effect on the properties and structure of Portland slag cement. *Cement & Concrete Composites* **30**: 679–685.

Lea FM, Hewlett PC and Liska M (2019) *Lea's chemistry of cement and concrete* (Hewlett PC and Liska M (eds)). Butterworth-Heinemann, Oxford, UK.

Lei Z, Pavia S and Wang X (2024) Biomass ash waste from agricultural residues: characterisation, reactivity and potential to develop one-part geopolymer cement. *Construction and Building Materials* **431**: 136544.

Leung CKY (2001) Concrete as a building material. In *Encyclopedia of Materials: Science and Technology* (Buschow KHJ, Cahn RW, Flemings MC *et al.* (eds)). Elsevier, Oxford, UK.

Li DX, Chen YM, Shen JL, Su JH and Wu XQ (2000) The influence of alkalinity on activation and microstructure of fly ash. *Cement and Concrete Research* **30**: 881–886.

Li GY and Zhao XH (2003) Properties of concrete incorporating fly ash and ground granulated blast-furnace slag. *Cement & Concrete Composites* **25**: 293–299.

Li GY (2004) Properties of high-volume fly ash concrete incorporating nano-SiO_2. *Cement and Concrete Research* **34**: 1043–1049.

Lin IJ and Somasundaran P (1972) Alterations in properties of samples during their preparation by grinding. *Powder Technology* **6**: 171–179.

Liska M and Al-Tabbaa A (2012) Performance of magnesia cements in porous blocks in acid and magnesium environments. *Advances in Cement Research* **24**: 221–232.

Liska M, Wilson A and Bensted J (2019) Special cements. In *Lea's Chemistry of Cement and Concrete*, 5th edn (Hewlett PC (ed.)). Butterworth-Heinemann, Oxford, UK.

Liu S, Wang L, Gao Y, Yu B and Bai Y (2014) Comparing study on hydration properties of various cementitious systems. *Journal of Thermal Analysis and Calorimetry* **118**: 1483–1492.

Low Carbon Concrete Group (2022) *Low Carbon Concrete Routemap*. Institution of Civil Engineers (ICE), London, UK.

Lu D and Wei S (1992) Effect of grinding aids on producing ultrafine particles. *Advanced Powder Technology* **3**: 47–53.

Lu D, Shi X and Zhong J (2022) Interfacial bonding between graphene oxide coated carbon nanotube fiber and cement paste matrix. *Cement and Concrete Composites* **134**: 104802.

Luo Z, Li, W, Wang K, Castel A and Shah SP (2021) Comparison on the properties of ITZs in fly ash-based geopolymer and Portland cement concretes with equivalent flowability. *Cement and Concrete Research* **143**: 106392.

Luukkonen T, Abdollahnejad Z, Yliniemi J, Kinnunen P and Illikainen M (2018) One-part alkali-activated materials: A review. *Cement and Concrete Research* **103**: 21–34.

Ma H, Cui C, Ma W, Yang Y and Tang G (2016) Enhancement mechanism of new type autoclaved calcium carbide residue shell-aggregate on concrete. *Cement and Concrete Composites* **72**: 146–154.

Martirena Hernández JF, Middendorf B, Gehrke M and Budelmann H (1998) Use of wastes of the sugar industry as pozzolana in lime-pozzolana binders: study of the reaction. *Cement and Concrete Research* **28**: 1525–1536.

Marwan T, Pera J and Ambroise J (1992) The action of some aggressive solutions on Portland and calcined laterite blended cement concrete. *Proceedings of the 4th International Conference on Fly Ash, Silica Fume, Slag and Natural Pozzolan in Concrete*, v.1, Istanbul 3–8 May 1992.

Meddah MS, Zitouni S and Belâabes S (2010) Effect of content and particle size distribution of coarse aggregate on the compressive strength of concrete. *Construction and Building Materials* **24**: 505–512.

Mehta P and Aïtcin P-C (1990) Principles underlying production of high-performance concrete. *Cement, Concrete and Aggregates* **12**: 70–78.

Mehta PK and Monteiro PJM (2014) *Concrete: Microstructure, Properties, and Materials*. McGraw-Hill Education, New York, NY, USA.

Mindess S (2019) Resistance of concrete to destructive agencies. In *Lea's Chemistry of Cement and Concrete*, 5th edn (Hewlett PC and Liska M (eds)). Butterworth-Heinemann, Oxford, UK.

Mo L and Panesar DK (2012) Effects of accelerated carbonation on the microstructure of Portland cement pastes containing reactive MgO. *Cement and Concrete Research* **42**: 769–777.

Monte MC, Fuente E, Blanco A and Negro C (2009) Waste management from pulp and paper production in the European Union. *Waste Management* **29**: 293–308.

Mordor Intelligence (2024) *Calcium Carbide Market Size & Share Analysis – Growth Trends & Forecasts (2024–2029)*. https://www.mordorintelligence.com/industry-reports/calcium-carbide-market (accessed 21/11/2024).

Mozaffari E, Kinuthia JM, Bai J and Wild S (2009) An investigation into the strength development of wastepaper sludge ash blended with ground granulated blastfurnace slag. *Cement and Concrete Research* **39**: 942–949.

MPA Precast and MPA Masonry (2024) *Precast Concrete and Masonry Sustainability Report*. MPA Precast and MPA Masonry, London, UK.

MPA the Concrete Centre (2017) *Specifying Sustainable Concrete Understanding the Role of Constituent Materials – TCC/05/24*. MPA the Concrete Centre, London, UK.

MPA the Concrete Centre (2019) *Concrete and Fire Safety: Guidance on the Use of Concrete and Masonry for Fire Resistant and Efficient Structures – TCC/03/62*. MPA the Concrete Centre, London, UK.

MPA the Concrete Centre (2020) *Specifying Sustainable Concrete Understanding the Role of Constituent Materials – TCC/05/24*. MPA the Concrete Centre, London, UK.

MPA UK Concrete (2020) *UK Concrete and Cement Industry Roadmap to Beyond Net Zero*. Mineral Products Association, London, UK.

Müllauer W, Beddoe RE and Heinz D (2015) Leaching behaviour of major and trace elements from concrete: effect of fly ash and GGBS. *Cement and Concrete Composites* **58**: 129–139.

Nanni A (1988) Curing of roller compacted concrete: strength development. *Journal of Transportation Engineering* **114**: 684–694.

National Grid (2023) *National Grid Completes Record-Breaking Pour of Cement-Free Concrete at London Power Tunnels*. National Grid, London, UK.

National Highways (2021) *Design Manual for Roads and Bridges – CD226 Design for New Pavement Construction Version 0.1.0*. National Highways, Department for Transport, London, UK.

Neville AM (1996) *Properties of Concrete*, 5th edn. Pearson, Cambridge, UK.

Ngo T, Mendis P and Krauthammer T (2007) Behavior of ultrahigh-strength prestressed concrete panels subjected to blast loading. *Journal of Structural Engineering* **133**: 1582–1590.

NSW Government Resources (2024) Industrial Mineral Fact Sheets – Olivine. https://www.resources.nsw.gov.au/sites/default/files/2022-11/olivine.pdf (accessed 26/02/2025).

Nuaklong P, Sata V and Chindaprasirt P (2016) Influence of recycled aggregate on fly ash geopolymer concrete properties. *Journal of Cleaner Production* **112**: 2300–2307.

Oderji SY, Chen B, Shakya C, Ahmad MR and Shah SFA (2019) Influence of superplasticizers and retarders on the workability and strength of one-part alkali-activated fly ash/slag binders cured at room temperature. *Construction and Building Materials* **229**: 116891.

Okoye FN, Durgaprasad J and Singh NB (2016) Effect of silica fume on the mechanical properties of fly ash based-geopolymer concrete. *Ceramics International* **42**: 3000–3006.

O'Rourke B, McNally C and Richardson MG (2009) Development of calcium sulfate-GGBS-Portland cement binders. *Construction and Building Materials* **23**: 340–346.

Orr J, Gibbons O and Arnold W (2022) *How to calculate embodied carbon*, 2nd edn. Institution of Structural Engineers, London, UK.

Osborne GJ (1999) Durability of Portland blast-furnace slag cement concrete. *Cement and Concrete Composites* **21**: 11–21.

Patil S, Somasekharaiah HM, Rao HS and Ghorpade VG (2022) Durability and micro-structure studies on fly ash and silica fume based composite fiber reinforced high-performance concrete. *Materials Today: Proceedings* **49**: 1511–1520.

Payá J, Monzó J, Borrachero MV and Peris-Mora E (1995) Mechanical treatment of fly ashes. Part I: physico-chemical characterization of ground fly ashes. *Cement and Concrete Research* **25**: 1469–1479.

PCA (Portland Cement Association) (2002) *Concrete Information – Types and Causes of Concrete Deterioration*. PCA, Illinois, IL, USA.

Persson B (1997) Self-desiccation and its importance in concrete technology. *Materials and Structures* **30**: 293–305.

Pigeon M and Pleau R (2010) *Durability of Concrete in Cold Climates*. Taylor and Francis, Abingdon, UK.

Plank J, Schroefl C, Gruber M, Lesti M and Sieber R (2009) Effectiveness of polycarboxylate superplasticizers in ultra-high strength concrete: the importance of PCE compatibility with silica fume. *Journal of Advanced Concrete Technology* **7**: 5–12.

Possan E, Felix EF and Thomaz WA (2016) CO_2 uptake by carbonation of concrete during life cycle of building structures. *Journal of Building Pathology and Rehabilitation* **1**: 7.

Prokopski G and Langier B (2000) Effect of water/cement ratio and silica fume addition on the fracture toughness and morphology of fractured surfaces of gravel concretes. *Cement and Concrete Research* **30**: 1427–1433.

Provis JL and Van Deventer JS (2013) *Alkali activated materials*. RILEM state-of-the-art report, TC 224-AAM. Springer Science and Business Media, New York, NY, USA.

Provis JL, Palomo A and Shi C (2015) Advances in understanding alkali-activated materials. *Cement and Concrete Research* **78**: 110–125.

Pu B-C, Liu B, Li L, Jiang L, Zhou J and Ding P (2024) Using rice husk ash in alkali-activated ultra-high-performance concrete: flowability, early age strength and elasticity modulus. *Construction and Building Materials* **443**: 137771.

Putranto A, Lin T-H and Huang B-X (2024) Deep learning-integrated electromagnetic imaging for evaluating reinforced concrete structures in water-contact scenarios. *Automation in Construction* **164**: 105459.

Qian JS, Shi CJ and Wang Z (2001) Activation of blended cements containing fly ash. *Cement and Concrete Research* **31**: 1121–1127.

Qiao J, Lv Y, Peng H, Han W, Pan B and Zhang B (2024) Performance and characterization of fly ash-metakaolin-based geopolymer pastes. *Materials Today Communications* **41**: 110293.

Radlinski M and Olek J (2012) Investigation into the synergistic effects in ternary cementitious systems containing Portland cement, fly ash and silica fume. *Cement and Concrete Composites* **34**: 451–459.

Rattanashotinunt C, Thairit P, Tangchirapat W and Jaturapitakkul C (2013) Use of calcium carbide residue and bagasse ash mixtures as a new cementitious material in concrete. *Materials & Design (1980–2015)* **46**: 106–111.

Roy A and Sadiqul Islam GM (2024) Geopolymer using different size fractions of recycled brick-based mixed demolition waste. *Cleaner Materials* **11**: 100224.

Rukzon S, Chindaprasirt P and Mahachai R (2009) Effect of grinding on chemical and physical properties of rice husk ash. *International Journal of Minerals, Metallurgy and Materials* **16**: 242–247.

Sadique M and Al-Nageim H (2012) Hydration kinetics of a low carbon cementitious material produced by physico-chemical activation of high calcium fly ash. *Journal of Advanced Concrete Technology* **10**: 254–263.

Sadique M and Coakley E (2016) The influence of physico-chemical properties of fly ash and CKD on strength generation of high-volume fly ash concrete. *Advances in Cement Research* **28**: 595–605.

Sadique M, Al-Nageim H, Atherton W, Seton L and Dempster N (2012) A new composite cementitious material for construction. *Construction and Building Materials* **35**: 846–855.

Sadique M, Al-Nageim H, Atherton W, Seton L and Dempster N (2013) Mechano-chemical activation of high-Ca fly ash by cement free blending and gypsum aided grinding. *Construction and Building Materials* **43**: 480–489.

Sadique M, Al-Nageim H, Atherton W, Seton L and Dempster N (2019) Analytical investigation of hydration mechanism of a non-Portland binder with waste paper sludge ash. *Construction and Building Materials* **211**: 80–87.

San Nicolas R, Bernal SA, Mejía De Gutiérrez R, Van Deventer JSJ and Provis JL (2014) Distinctive microstructural features of aged sodium silicate-activated slag concretes. *Cement and Concrete Research* **65**: 41–51.

Sarkar SL, Aimin X and Jana D (2001) Scanning electron microscopy, X-ray microanalysis of concretes. In *Handbook of Analytical Techniques in Concrete Science and Technology* (Ramachandran VS and James JB (eds)). William Andrew Publishing, Norwich, NY, USA.

Sarkar K, Shiuly A and Dhal KG (2024) Revolutionizing concrete analysis: An in-depth survey of AI-powered insights with image-centric approaches on comprehensive quality control, advanced crack detection and concrete property exploration. *Construction and Building Materials* **411**: 134212.

Segui P, Aubert JE, Husson B and Measson M (2012) Characterization of wastepaper sludge ash for its valorization as a component of hydraulic binders. *Applied Clay Science* **57**: 79–85.

Sekulic Z, Petrov M and Zivanovic D (2004) Mechanical activation of various cements. *International Journal of Mineral Processing* **74**, Supplement: S355–S363.

Sekulic Z, Popov S, Duricic M and Rosic A (1999) Mechanical activation of cement with addition of fly ash. *Materials Letters* **39**: 115–121.

Senneca O, Salatino P, Chirone R, Cortese L and Solimene R (2011) Mechanochemical activation of high-carbon fly ash for enhanced carbon reburning. *Proceedings of the Combustion Institute* **33**: 2743–2753.

Shah V and Scott A (2022) Pozzolanic characteristics of silica recovered from olivine. *Construction and Building Materials* **332**: 127378.

Shamsaei E, De Souza FB, Yao X *et al.* (2018) Graphene-based nanosheets for stronger and more durable concrete: a review. *Construction and Building Materials* **183**: 642–660.

Shanks B, Howe C, Draper S, Wong H and Cheeseman C (2024) Production of low-carbon amorphous SiO_2 for use as a supplementary cementitious material and nesquehonite from olivine. *Materials Letters* **361**: 136133.

Shanks W, Dunant CF, Drewniok MP *et al.* (2019) How much cement can we do without? Lessons from cement material flows in the UK. *Resources, Conservation and Recycling* **141**: 441–454.

Shi C and Xie P (1998) Interface between cement paste and quartz sand in alkali-activated slag mortars. *Cement and Concrete Research* **28**: 887–896.

Shi C, Tu Z, Guo M-Z and Wang D (2017) Accelerated carbonation as a fast curing technology for concrete blocks. In *Sustainable and Nonconventional Construction Materials using Inorganic Bonded Fiber Composites* (Savastano Junior H, Fiorelli J and Dos Santos SF (eds)). Woodhead Publishing, Cambridge, UK.

Shoaei P, Momenzadeh A, Hosseini H *et al.* (2024) One-part slag/zeolite geopolymer mortars under ambient and heat curing conditions. *Case Studies in Construction Materials* **20**: e02677.

Sidhu AS and Siddique R (2024) Review on effect of curing methods on high strength concrete. *Construction and Building Materials* **438**: 136858.

Singh B, Ishwarya G, Gupta M and Bhattacharyya SK (2015a) Geopolymer concrete: A review of some recent developments. *Construction and Building Materials* **85**: 78–90.

Singh SB, Munjal P and Thammishetti N (2015b) Role of water/cement ratio on strength development of cement mortar. *Journal of Building Engineering* **4**: 94–100.

Singh VK (2023) Hydration and setting of Portland cement. In *The Science and Technology of Cement and Other Hydraulic Binders* (Singh VK (ed.)). Woodhead Publishing, Cambridge, UK.

Sokhansefat G, Ley MT, Cook MD, Alturki R and Moradian M (2019) Investigation of concrete workability through characterization of aggregate gradation in hardened concrete using X-ray computed tomography. *Cement and Concrete Composites* **98**: 150–161.

Somasri M and Narendra Kumar B (2021) Graphene oxide as nano material in high strength self-compacting concrete. *Materials Today: Proceedings* **43**: 2280–2289.

Soutsos M and Domone P (eds) (2017) *Construction Materials: Their Nature and Behaviour*, 5th edn. CRC Press, Boca Raton, FL, USA.

Standing Committee on Structural Safety (SCOSS) (2018) Fire in Multi-Storey Car Parks. https://www.cross-safety.org/uk/safety-information/cross-safety-alert/fire-multi-storey-car-parks (accessed 26/02/2025).

Stopic S, Dertmann C, Koiwa I *et al.* (2019) Synthesis of nanosilica via olivine mineral carbonation under high pressure in an autoclave. *Metals* **9**: 708.

Swathi B and Vidjeapriya R (2023) Synergistic use of nano-silica to enhance the characterization of ambient-cured geopolymer concrete. *Archives of Civil and Mechanical Engineering* **24**: 4.

Tan KF and Pu XC (1998) Strengthening effects of finely ground fly ash, granulated blast furnace slag, and their combination. *Cement and Concrete Research* **28**: 1819–1825.

Taylor P and Sadati S (2021) *Reduced Cementitious Material in Optimized Concrete Mixtures (No. NRRA202102)*. Minnesota Department of Transportation, Willmar, MN, USA.

Thakkar S, Chanda A and Rawat S (2023) Early strength development of self-compacting geopolymer mortar using binary and tertiary blends of industrial wastes. *Materials Today: Proceedings* **93**: 467–474.

The Concrete Society (2024) *Concrete @ your Fingertips – High strength concrete.* https://www.concrete.org.uk/fingertips-nuggets.asp?cmd=display&id=528 (accessed 11/10/2024).

Tian W, Qi B, Liu Y, Liu K and Wang W (2021) Early frost resistance and permeability properties of carbon fiber/cement-based composite cured by ohmic heating under ultra-low temperature. *Construction and Building Materials* **282**: 122729.

Torres De Sande V, Sadique M, Bras A and Pineda P (2022) Activated sugarcane bagasse ash as efficient admixture in cement-based mortars: mechanical and durability improvements. *Journal of Building Engineering* **59**: 105082.

Toutanji HA and Elkorchi T (1995) The influence of silica fume on the compressive strength of cement paste and mortar. *Cement and Concrete Research* **25**: 1591–1602.

Tunstall LE, Ley MT and Scherer GW (2021) Air entraining admixtures: Mechanisms, evaluations, and interactions. *Cement and Concrete Research* **150**: 106557.

UKQAA (United Kingdom Quality Ash Association) (2010) *Annual PFA Statistics*. UKQAA, Royal Sutton Coldfield, UK.

Van Deventer JSJ, Provis JL and Duxson P (2012) Technical and commercial progress in the adoption of geopolymer cement. *Minerals Engineering* **29**: 89–104.

Verbeck G and Landgren R (1960) Influence of physical characteristics of aggregates on frost resistance of concrete. *Proceedings of American Society for Testing and Materials* **60**.

Viswanath HR, Tolloczko J and Clarke JN (1998) *Multi-purpose High-rise Towers and Tall Buildings*. CRC Press, Boca Raton, FL, USA.

Wang BM and Wang LJ (2004) *Development of Studies and Applications of Activation Techniques of Fly Ash*. Tsunghua University, Beijing, China.

Wang K, Mishulovich A and Shah SP (2007) Activations and properties of cementitious materials made with cement-kiln dust and Class F fly ash. *Journal of Materials in Civil Engineering* **19**: 112–119.

Wang C, Li K, Li H and Wang J (2008) Surface thermal treatment of carbon fibers and electromagnetic shielding of carbon fiber/cement composites. *Kuei Suan Jen Hsueh Pao/Journal of the Chinese Ceramic Society* **36**: 1348–1355.

Wang F, Dreisinger D, Jarvis M and Hitchins T (2019) Kinetics and mechanism of mineral carbonation of olivine for CO_2 sequestration. *Minerals Engineering* **131**: 185–197.

Wang Y, Zhang W, Wang J *et al.* (2024a) Effects of coarse aggregate size on thickness and microproperties of ITZ and the mechanical properties of concrete. *Cement and Concrete Composites* **154**: 105777.

Wang Z, Bai E, Liu C, Ren, B and Zhang J (2024b) Strength, toughness and interface modification of graphene oxide grafted carbon fiber modified concrete. *Construction and Building Materials* **428**: 136336.

Waste and Resources Action Programme (2007) *A New Approach to Paper Mill Sludge*. WRAP, Banbury, UK.

Winnefeld F, Leemann A, Lucuk M, Svoboda P and Neuroth M (2010) Assessment of phase formation in alkali activated low and high calcium fly ashes in building materials. *Construction and Building Materials* **24**: 1086–1093.

Wu Z, Shi C and Khayat KH (2016) Influence of silica fume content on microstructure development and bond to steel fiber in ultra-high strength cement-based materials (UHSC). *Cement and Concrete Composites* **71**: 97–109.

Wu C, Li J and Su Y (2018) Introduction. In *Development of Ultra-High Performance Concrete Against Blasts* (Wu C, Li J and Su Y (eds)). Woodhead Publishing, Cambridge, UK.

Wu Q, Xue Q and Yu Z (2021) Research status of super sulfate cement. *Journal of Cleaner Production* **294**: 126228.

Wu D, Xiong X, Ma, J, Lv L and Huang J (2023) Enhanced thermoelectric and compressive performance of MnO_2 coated carbon fiber/cement composites by microwave electrodeposition/ microwave alkali treatment. *Ceramics International* **49**: 37535–37548.

Wudil YS, Shalabi AF, Al-Osta MA, Gondal MA and Al-Nahari E (2024) Effective corrosion detection in reinforced concrete via laser-induced breakdown spectroscopy and machine learning. *Materials Today Communications* **41**: 111005.

www.3ders.org (2014) *10 completely 3D printed houses appears in Shanghai, built under a day.* https://www.3ders.org/articles/20140401-10-completely-3d-printed-houses-appears-in-shanghai-built-in-a-day.html (accessed 25/09/2024).

Xian X, Zhang D, Lin H and Shao Y (2022) Ambient pressure carbonation curing of reinforced concrete for CO_2 utilization and corrosion resistance. *Journal of CO_2 Utilization* **56**: 101861.

Xu G and Beaudoin JJ (2000) Effect of polycarboxylate superplasticizer on contribution of interfacial transition zone to electrical conductivity of Portland cement mortars. *ACI Materials Journal* **97**: 418–424.

Xu Z and Gao C (2015) Graphene fiber: a new trend in carbon fibers. *Materials Today* **18**: 480–492.

Yang K-H, Cho A-R, Song J-K and Nam S-H (2012) Hydration products and strength development of calcium hydroxide-based alkali-activated slag mortars. *Construction and Building Materials* **29**: 410–419.

Yang R, Yu R, Shui Z *et al.* (2019) Low carbon design of an ultra-high performance concrete (UHPC) incorporating phosphorous slag. *Journal of Cleaner Production* **240**: 118157.

Yazdi M and Sarafrazi KJC (2014) Concrete, automated segmentation of concrete images into microstructures: a comparative study *Computers and Concrete* **14**: 315–325.

Yazici H (2008) The effect of silica fume and high-volume Class C fly ash on mechanical properties, chloride penetration and freeze–thaw resistance of self-compacting concrete. *Construction and Building Materials* **22**: 456–462.

Yu R, Spiesz P and Brouwers HJH (2015) Development of an eco-friendly ultra-high performance concrete (UHPC) with efficient cement and mineral admixtures uses. *Cement and Concrete Composites* **55**: 383–394.

Zhang D, Li S, Song W and Lin W (2016) Cyclic CO_2 capture performance of carbide slag. *Energy Sources, Part A: Recovery, Utilization, and Environmental Effects* **38**: 577–582.

Zhang H, Ma W, Gao F *et al.* (2024) Rheology, shrinkage, mechanical properties and microstructure of ultra-light-weight concrete with fly ash cenospheres. *Journal of Building Engineering* **98**: 111258.

Zhao H, Sun W, Wu X and Gao B (2012) The effect of coarse aggregate gradation on the properties of self-compacting concrete. *Materials & Design* **40**: 109–116.

Zheng S, Zhou X, Xing W and Zhao P (2022) Analysis on the evolution characteristics of kaolin international trade pattern based on complex networks. *Resources Policy* **77**: 102783.

Zhou D, Wang R, Tyrer M, Wong H and Cheeseman C (2017) Sustainable infrastructure development through use of calcined excavated waste clay as a supplementary cementitious material. *Journal of Cleaner Production* **168**: 1180–1192.

Zhu Z, Zhang Q, Zhang H *et al.* (2023) Effect of different superplasticizers on the mechanism, workability, and microstructure of biomass-activated grouts. *Construction and Building Materials* **373**: 130857.

Monower Sadique
ISBN 978-1-83608-477-8
https://doi.org/10.1108/978-1-83608-474-720251009

Index

Printed and bound by CPI Group (UK) Ltd, Croydon, CR0 4YY
02/07/2025
14698051-0001